THE ENDING OF TIME

Also by
Jiddu Krishnamurti

The Ending
of Time

WHERE PHILOSOPHY & PHYSICS MEET

Revised and Expanded Edition

JIDDU KRISHNAMURTI
AND DAVID BOHM

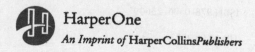

HarperOne
An Imprint of HarperCollinsPublishers

HarperOne

Contents

Introduction

These dialogues between Jiddu Krishnamurti and the theoretical physicist David Bohm started by addressing the origin of human conflict. Both men agreed in attributing this to the separative and time-bound nature of the self and the way that it conditions us to rely wrongly on thought, which is based on inevitably limited past experience. The possibility of insight that will end this flawed mentality was discussed in depth. The focus then shifted to an inquiry into the significance of death, and to a discussion probing the "ground" of being and the place of consciousness in the universe. The final dialogues reviewed the profound linkage that Krishnamurti and Bohm saw between these ultimate questions and everyday life, and what we can do about the barriers that stand in the way.

The backgrounds of the two men could hardly be more different. Born in India, Jiddu Krishnamurti was chosen by the Theosophical Society at the age of thirteen to be a "vehicle" for the World Teacher, a role which he firmly renounced at the age of thirty-four. Without any formal education, he then travelled the world giving talks and interviews until the age of ninety. Rejecting any kind of professional title for himself, and even any kind of formal description of his talks, he spoke to his audiences "as a friend" and disclaimed any authority, urging listeners to test the truth of his words in their daily lives. Meditation and its insights were for Krishnamurti the way of life.

Born in the United States, David Bohm, one of the outstanding theoretical physicists of the twentieth century, graduated from Pennsylvania State College and obtained at the age of twenty-six his Ph.D. in physics at the University of California, Berkeley, under the direction of Robert Oppenheimer. He then taught at Princeton, working closely with Einstein. Because of blacklisting, at the time of the McCarthy era, for alleged pro-communist sympathies, he was forced to leave the United States and took up a post at the University of São Paulo in Brazil. He completed his career as professor of theoretical physics at Birkbeck College, University of London. He was a prolific author of works on physics and was engaged in path-breaking research until his death.

This book has been prepared from dialogues that took place between Jiddu Krishnamurti and Professor David Bohm in America and in England between April and September 1980. On certain occasions other people were present, and their occasional contributions to the discussions, unless otherwise stated, are attributed to "Questioner" rather than to individuals by name.

The appendix that forms the second part of the book consists of two conversations in 1983 between Bohm and Krishnamurti on the future of humanity. Although the conversations contain this new and different topic, they have been included in the book because they explore exchanges that develop some crucial points in the "Ending of Time" dialogues.

—DAVID SKITT, 2014

The Ending
of Time

Foreword

In March 2013 a striking map appeared on the world's TV screens of the state of the universe 380,000 years after the Big Bang. Public interest at this "infant" image flared but quite quickly subsided, perhaps because after decades of human exploration of space, we now take these findings for granted. The Hubble Space Telescope has for years been sending back astonishing, often beautiful portraits of distant galaxies, showing massive convulsions of energy billions of light-years away.

Now that these extraordinary vistas are being revealed to us, it is perhaps not surprising that a number of philosophers and scientists have taken up the ultimate issue of the place of human consciousness in the universe. What we can see and learn from the cosmos clearly raises challenging and daunting questions. Long ago Blaise Pascal, the French scientist and religious writer, found infinite space "frightening," and in our time the evolutionary biologist Richard Dawkins assures us that the universe "doesn't care" about human preferences—an issue that Jiddu Krishnamurti and the physicist David Bohm discussed in *The Ending of Time*. The vast spans of time and distance of the universe also seem to make almost footling any deep probing of it by minds on a small "third rock from the sun."

The philosopher Thomas Nagel summed up well in his book *Mind and Cosmos* the modesty called for in dealing with this subject but also the case for ongoing inquiry:

> I would like to extend the boundaries of what is not regarded as unthinkable, in light of how little we really understand about the world. . . . It is perfectly possible that the truth is beyond our reach, in virtue of our intrinsic cognitive limitations, and not merely beyond our grasp in humanity's present stage of intellectual development. But I believe that we cannot know this, and that it makes sense to go on seeking a systematic understanding of how we and other living beings fit into the world.[1]

And the neuroscientist Christof Koch, in his book *Consciousness: Confessions of a Romantic Reductionist*, boldly argued that

> the entire cosmos is suffused with sentience. We are surrounded and immersed in consciousness; it is in the air we breathe, the soil we tread on, the bacteria that colonize our intestines, and the brain that enables us to think.[2]

---

1. Thomas Nagel, *Mind and Cosmos: Why the Materialist Neo-Darwinian Conception of Nature Is Almost Certainly False* (New York: Oxford University Press, 2012).

2. Christof Koch, *Consciousness: Confessions of a Romantic Reductionist* (Cambridge, MA: MIT Press, 2012).

Both Thomas Nagel and Christof Koch reflected in their views a willingness to expand, each in his own way, the horizons of the human mind, and it was in this fresh climate of innovative inquiry that the "Ending of Time" dialogues between Jiddu Krishnamurti and David Bohm fell naturally into place. Krishnamurti, for his part, made the psychological case for the illusions of the separative self needing to end before the mind can delve deeply into issues such as the human place in the cosmos. And he and Bohm discussed the cause of these illusions, and particularly the conflict they engender, in detail. But he also went on to argue strongly that rational inquiry can clear the way for profound and transforming insight into this question. Does such deep exploration have implications for our everyday lives? Krishnamurti argued that it does; the human being does not have to be a timorous witness, an awed or baffled outcast in the universe. What the ending of such "psychological exile" might mean for humanity was explored in depth.

In this new enlarged edition, two more dialogues have been added to the thirteen in the previous book. This complete edition also contains, in an appendix, two later dialogues in a part entitled "The Future of Humanity," which David Bohm described in a foreword as both amplifying and possibly serving as an introduction to those in *The Ending of Time*. The editors felt, however, that the distinction made between brain and mind in the second of these dialogues could most usefully conclude the whole series. Readers are invited to make their own choice.

Study of these fifteen dialogues will also benefit from combining the reading of this book with listening to the audio recordings of them. There are often pauses in these recordings to allow pondering of what is said, at times also touches of humour, and careful and constant probing by both Jiddu Krishnamurti and David Bohm to find the best words for their exchanges. Please see www .jkrishnamurti.org for a selection of audio recordings available for download.

—DAVID SKITT

One

The Roots of Psychological Conflict

1 APRIL 1980, OJAI, CALIFORNIA

JIDDU KRISHNAMURTI: How shall we start? I would like to ask if humanity has taken a wrong turn.

DAVID BOHM: A wrong turn? Well, it must have done so, a long time ago, I think.

JK: That is what I feel. A long time ago. It appears that way. Why? You see, as I look at it, I am just enquiring, mankind has always tried to *become* something.

DB: Well, possibly. I was struck by something I once read about man going wrong about five or six thousand years ago, when he began to be able to plunder and take slaves. After that, his main purpose of existence was just to exploit and plunder.

JK: Yes, but there is the sense of inward becoming.

DB: Well, we should make it clear how this is connected. What kind of becoming was involved in doing that? Instead of being constructive, and discovering new techniques and tools and so on, man at a certain time found it easier to plunder his neighbours. Now, what did they want to become?

JK: Conflict has been the root of all this.

DB: But what was the conflict? If we could put ourselves in the place of those people of long ago, how would you see that conflict?

JK: What is the *root* of conflict? Not only outwardly but also of this tremendous inward conflict of humanity. What is the *root* of it?

DB: Well, it seems that it is contradictory desires.

JK: No. Is it that in all religions you must become something? You must reach something?

DB: Then what made people want to do that? Why weren't they satisfied to be whatever they were? You see, the religion would not have caught on unless people felt that there was some attraction in becoming something more.

JK: Isn't it an avoidance, not being able to face the fact and change the fact, but rather moving to something else—to more and more and more?

DB: What would you say was the fact that people couldn't stay with?

JK: What was the fact? The Christians have said there was Original Sin.

DB: But the wrong turn happened long before that.

JK: Yes, long before that. Long before that, the Hindus had this idea of karma. So what is the origin of all this?

DB: We have said that there was the fact that people couldn't stay with. Whatever it was, they wanted to imagine something better.

JK: Yes, something better. Becoming more and more.

DB: And you could say that they began to make things technologically better, then they extended this and said, "I too must become better."

JK: Yes, inwardly become better.

DB: All of us together must become better.

JK: That's right. But what is the root of all this?

DB: Well, I should think it is natural in thought to project this goal of becoming better. That is, it is intrinsic in the structure of thought.

JK: Is it that the principle of becoming better outwardly has moved to becoming better inwardly?

DB: If it is good to become better outwardly, then why shouldn't I become better inwardly?

JK: Is that the cause of it?

DB: That is getting towards it. It's coming nearer.

JK: Is it coming nearer? Is time the factor? Time as "I need knowledge in order to do this or that"? The same principle applied inwardly? Is time the factor?

DB: I can't see that time by itself can be the only factor.

JK: No, no. Time. Becoming—which implies time.

DB: Yes, but we don't see how time is going to cause trouble. We have to say that time applied outwardly doesn't cause any difficulty.

JK: It causes a certain amount—but we are discussing the idea of time inwardly.

DB: So we have to see why time is so destructive inwardly.

JK: Because I am trying to become something.

DB: Yes, but most people would say that this is only natural. You have to explain what it is that is wrong about becoming.

JK: Obviously, there is conflict, in that when I am trying to become something, it is a constant battle.

DB: Yes. Can we go into that: Why is it a constant battle? It is not a battle if I try to improve my position outwardly.

JK: Outwardly no. It is more or less all right outwardly, but when that same principle is applied inwardly it brings about a contradiction.

DB: And the contradiction is . . . ?

JK: Between "what is" and "becoming what should be."

DB: The difficulty is why is it a contradiction inwardly and not outwardly?

JK: Inwardly it builds up a centre, doesn't it? An egotistic centre?

DB: Yes, but can we find some reason why it should do so? Does it build up when we do it outwardly? It seems it need not.

JK: It need not.

DB: But when we are doing it inwardly, then we are trying to force ourselves to be something that we are not. And that is a fight.

JK: Yes. That is a fact. Is it that one's brain is so accustomed to conflict that one rejects any other form of living?

DB: After a while people come to the conclusion that conflict is inevitable and necessary.

JK: But what is the origin of conflict?

DB: I think we touched on that by saying that we are trying to force ourselves. When we are a certain thing that we want to be, we also want to be something else that is different; and therefore we want two different things at the same time. Would that seem right?

JK: I understand that. But I am trying to find out the origin of all this misery, confusion, conflict, struggle—what is the beginning of it? That's why I asked at the beginning: Has mankind taken a wrong turn? Is the origin "I" and "not I"?

DB: I think that is getting closer. The separation between "I" and "not I."

JK: Yes, that's it. And the "I"—why has mankind created this "I," which must, inevitably, cause conflict? "I" and "you," and "I" better than "you," and so on and so on.

DB: I think it was a mistake made a long time ago, or, as you call it, a wrong turn, that having introduced separation between various things outwardly, we then kept on doing it—not out of ill will but simply through not knowing better.

JK: Quite.

DB: Not seeing what we were doing.

JK: Is that the origin of all this?

DB: I am not sure that it is the origin. What do you feel?

JK: I am inclined to observe that the origin is the ego, the "me," the "I."

DB: Yes.

JK: If there is no ego, there is no problem, there is no conflict, there is no time—time in the sense of becoming or not becoming, being or not being.

DB: But it might be that we would still slip into whatever it was that made us make the ego in the first place.

JK: Wait a minute. Is it that energy—being so vast, limitless—has been condensed or narrowed down in the mind, and the brain itself has become narrowed down because it couldn't contain all this enormous energy? You are following what I am saying?

DB: Yes.

JK: And therefore the brain has gradually narrowed down to "me," to the "I."

DB: I don't quite follow that. I understand that that is what happened, but I don't quite see all the steps. You say energy was enormous and the brain couldn't handle it, or decided that it couldn't handle it?

JK: It couldn't handle it.

DB: But if it can't handle it, it seems as if there is no way out.

JK: No. Just a minute. Go slowly. I just want to enquire, push into it a little bit. Why has the brain, with all thought, created this sense of "me," "I"? Why?

DB: We needed a certain sense of identity to function. Outwardly it had to be that way.

JK: Yes, to have a function.

DB: To know where we belong.

JK: Yes. And is that the movement which has brought the "me"? The movement of the outer? Where I had to identify with the family, the house, the trade or profession. All this gradually became the "me"?

DB: I think that this energy that you are talking about also entered into it.

JK: Yes, but I want to lead up to that slowly.

DB: You see, what you say is right, that in some way this sense of the "me" gradually strengthened, but by itself that wouldn't explain the tremendous strength that the ego has. It would only be a habit then. The ego becoming completely dominant required that it should become the focus of the greatest energy, of all the energy.

JK: Is that it? That the brain cannot hold this vast energy?

DB: Let's say that the brain is trying to control this—to bring it to order.

JK: Energy has no order.

DB: But if the brain feels it can't control something that is going on inside, it will try to establish order.

JK: Could we say that the brain, your brain, his brain, her brain, has not just been born; it is very, very old?

DB: In what sense?

JK: In the sense that it has evolved.

DB: Evolved, yes, from the animal. And the animal has evolved. So let's say that in a sense this whole evolution is somehow contained in the brain.

JK: I want to question evolution. I understand, say, evolution from the bullock cart to the jet.

DB: Yes. But before you question, we have to consider the evidence of man developing through a series of stages. You can't question that, can you?

JK: No, of course not.

DB: I mean, physically it is clear that evolution has occurred in some way.

JK: Physically, yes.

DB: And the brain has got larger, more complex. But you may question whether mentally evolution has any meaning.

JK: You see, I want to [*laughs*] abolish time, psychologically. You understand?

DB: Yes, I understand.

JK: To me that is the enemy. And is that the cause, the origin of man's misery?

DB: This use of time, certainly. Man had to use time for a certain purpose, but he misused it.

JK: I understand that. If I have to learn a language, I must have time.

DB: But the misuse of time by extending it inwardly...

JK: *Inwardly*. That is what I am talking about. Is that the cause of man's confusion—introducing time as a means of becoming, and becoming more and more perfect, more and more evolved, more and more loving? You follow what I mean?

DB: Yes, I understand. Certainly if we didn't do that, the whole structure would collapse.

JK: That's it.

DB: But I don't know whether there is not some other cause.

JK: Just a minute. I want to go into that a little bit. I am not talking theoretically, personally. But to me the idea of tomorrow doesn't exist psychologically—that is, time is a movement, either inwardly or outwardly.

DB: You mean psychological time?

JK: Yes, psychological time, and time outwardly. Now, if psychological time doesn't exist, then there is no conflict, there is no "me," no "I," which is the origin of conflict. Outwardly, technologically man has moved, evolved.

DB: And also in the inward physical structure.

JK: The structure, everything. But psychologically we have also moved outward.

DB: Yes, we have focused our life on the outward. Is that what you are saying?

JK: Yes. We have extended our capacities outwardly. And inwardly it is the same movement as outwardly. Now, if there is no inward movement as time, moving, becoming more and more, then what takes place? You understand what I am trying to convey? Time ends. You see, the outer movement is the same as the inward movement.

DB: Yes. It is going around and around.

JK: Involving time. If the movement ceases, then what takes place? I wonder if I am conveying anything? Could we put it this way? We have never touched any other movement than the outer movement.

DB: Generally anyway. We put most of our energy into the outward movement.

JK: And psychological movement is also outward.

DB: Well, it is the reflection of that outward movement.

JK: We think it is inward but it is actually outward, right?

DB: Yes.

JK: Now, if that movement ends, as it must, then is there a really inward movement—a movement not in terms of time?

DB: You are asking, is there another kind of movement which still moves, but not in terms of time?

JK: That's right.

DB: We have to go into that. Could you go further?

JK: You see, that word "movement" means time.

DB: Well, it really means to change from one place to another. But anyway there is still the notion of something which is not static. By denying time you don't want to return to something static, which is still time.

JK: Let's say, for instance, that one's brain has been trained, accustomed for centuries, to go north. And it suddenly realizes that going north means everlasting conflict. As it realizes that, the brain itself changes—the quality of the brain changes.

DB: All right. I can see it will wake up in some way to a different movement.

JK: Yes, different.

DB: Perhaps the word "flow" is better.

JK: I have been going north all my life, and there is a sudden stoppage from going north. But the brain is not going east or south or west. Then conflict ceases, right? Because it is not moving in any direction.

DB: So that is the key point—the direction of movement. When the movement is fixed in direction, inwardly, it will come to conflict. But outwardly we need a fixed direction.

JK: Of course we do. That's understood.

DB: Yes. So if we say the brain has no fixed direction, then what is it doing? Is it moving in all directions?

JK: I am a little bit hesitant to talk about this. Could one say, when one really comes to that state, that it is the source of all energy?

DB: Yes, as one goes deeper and more inward.

JK: This is the real inwardness; not the outward movement becoming the inner movement, but when there is no outer or inner movement...

DB: Yes, we can deny both the outward and the inner, so that all movement would seem to stop.

JK: Would that be the source of all energy?

DB: Yes, perhaps we could say that.

JK: May I talk about myself a little bit?

DB: Yes, go ahead.

JK: First about meditation. Conscious meditation is no meditation, right?

DB: What do you mean by conscious meditation?

JK: Deliberate, practised meditation, which is really premeditated meditation. Is there a meditation which is not premeditated—which is not the ego trying to become something—or the ego not trying to negate, negatively or positively?

DB: Before we go ahead, could we suggest what meditation should be? Is it an observation of the mind observing?

JK: No. It has gone beyond all that. I am using the word "meditation" in the sense in which there is not a particle of endeavour, of any sense of trying to become, consciously reach a level.

DB: The mind is simply with itself, silent.

JK: That is what I want to get at.

DB: It is not looking for anything.

JK: You see, I don't meditate in the normal sense of the word. What happens with me is—I am not talking personally, please—what happens with me is that I wake up meditating.

DB: In that state.

JK: One night at Rishi Valley in India I woke up. A series of incidents had taken place; there had been meditation for some days. It was a quarter past twelve; I looked at the watch [*laughs*]. And—I hesitate to say this, because it sounds extravagant and rather childish—the source of all energy had been reached. And that had an extraordinary effect on the brain, and also physically. Sorry to talk about myself, but you understand, literally any sense of . . . I don't know how to put it . . . any sense of the world and me, and that—you follow?—there was no division at all. Only this sense of tremendous source of energy.

DB: So the brain was in contact with this source of energy?

JK: Yes. Now, coming down to earth, and as I have been talking for sixty years, I would like another to reach this—no, not reach it. You understand what I am saying? Because all our problems—political, religious—all are resolved. Because it is pure energy from the very beginning of time. Now, how am I—please, not "I," you understand—how is one not to teach, not to help, or push, but how is one to say, "This way leads to a complete sense of peace, of love"? I am sorry to use all these words. But suppose you have come to that point and your brain itself is throbbing with it. How would you help another? You understand? Help—not words. How would you help another to come to that? You understand what I am trying to say?

DB: Yes.

JK: My brain—brain not mine—the brain has evolved. Evolution implies time, and it can only think, live in time. Now, for the brain to deny time is a tremendous activity, of having no problems, for any problem that arises, any question, is immediately solved.

DB: Is this situation sustained or is it only for a period?

JK: It is sustained, obviously, otherwise there is no point in it. It is not sporadic or intermittent. Now, how are you to open the door, how are you to help another to say, "Look, we have been going in the wrong direction, there is only non-movement; and, if movement stops, everything will be correct"?

DB: Well, it is hard to know beforehand if everything is going to be correct.

JK: Let's go back to what we began with. That is, has mankind taken a wrong turn, psychologically not physically? Can that turn be completely reversed? Or stopped? My brain is so accustomed to this evolutionary idea that I will become something, I will gain something, that I must have more knowledge and so on; can that brain suddenly realize that there is no such thing as time? You understand what I am trying to say?

DB: Yes.

JK: I was listening the other day to a discussion on television about Darwin, his voyage and what he achieved—his whole theory of evolution. It seems to me that this is totally untrue psychologically.

DB: It seems that he has given evidence that all species have changed in time. Why is that untrue?

JK: Yes, of course. It was obvious.

DB: It is true in that regard. I think it would be untrue to say the mind evolved in time.

JK: Of course.

DB: But physically it seems clear there has been a process of evolution, and that this has increased the capacity of the brain to do certain things. For example, we couldn't be discussing this if the brain had not grown larger.

JK: Of course.

DB: But I think you are implying that the mind is not originating in the brain. Is that so? The brain is perhaps an instrument of the mind?

JK: And the mind is not time. Just see what that means.

DB: The mind does not evolve with the brain.

JK: The mind not being of time, and the brain being of time—is that the origin of conflict?

DB: Well, we have to see why that produces conflict. It is not clear to say that the brain is of time, but rather that it has developed in such a way that time is in it.

JK: Yes, that is what I meant.

DB: But not necessarily so.

JK: It has evolved.

DB: It has evolved, so it has time within it.

JK: Yes, it has evolved, time is part of it.

DB: It has become part of its very structure.

JK: Yes.

DB: And that was necessary. However, the mind operates without time, although the brain is not able to do so.

JK: No. That means that God is in man, and God can only operate if the brain is quiet, if the brain is not caught in time.

DB: Well, I wasn't meaning that. I see that the brain, having a structure of time, is not able to respond properly to mind. That's really what seems to be involved here.

JK: Can the brain itself see that it is caught in time, and that as long as it is moving in that direction, conflict is eternal, endless? You follow what I am saying?

DB: Yes. Does the brain see it?

JK: Has the brain the capacity to see that what it is doing now—being caught in time—that in that process there is no end to conflict? That means, is there a part of the brain which is not of time?

DB: Not caught or functioning in time?

JK: Can one say that?

DB: I don't know.

JK: That would mean—we come back to the same thing in different words—that the brain is not being completely conditioned by time, so there is a part of the brain that is free of time.

DB: Not a part, but rather that the brain is mainly dominated by time, although that doesn't necessarily mean it couldn't shift.

JK: Yes. That is, can the brain, dominated by time, not be subservient to it?

DB: That's right. In that moment it comes out of time. I think I can see this; it is dominated only when you give it time. Thought which takes time is dominated, but anything fast enough is not dominated.

JK: Yes, that's right. Can the brain—which has been used to time—can it see in that process that there is no end to conflict? See in the sense of realizing this? Will it realize it under pressure? Certainly not. Will it realize it under coercion, reward, or punishment? It will not. It will either resist or escape.

So what is the factor that will make the brain see that the way it has been functioning is not correct? Let's use the word "correct" for the moment. And what will make it suddenly realize that that is totally mischievous? What will make it? Certainly not drugs or some kind of chemical.

DB: None of these outward things.

JK: Then what will make the brain realize this?

DB: What do you mean by realize?

JK: Realize that the path along which the brain has been going will always be the path of conflict.

DB: I think this raises the question that the brain resists such a realization.

JK: Of course, of course. Because it has been used to the old path, for centuries! How will you make the brain realize this fact? If you could make it realize that, conflict is finished.

You see, people have tried fasting, no sex, austerity, poverty, chastity in the real sense, purity, having a mind that is absolutely correct; they have tried going away by themselves; they have tried practically everything that man has invented, but none of these ways has succeeded.

DB: Well, what do you say? It is clear that people pursuing these outward goals are still becoming.

JK: Yes, but they never realize that these are outward goals. It means denying all that completely.

DB: You see, to go further, I think that one has to deny the very notion of time in the sense of looking forward to the future, and deny all the past.

JK: That's just it.

DB: That is, the whole of time.

JK: Time is the enemy [*laughs*]. Meet it, and go beyond it.

DB: Deny that it has an independent existence. You see, I think we have the impression that time exists independently of us. We are in the stream of time, and therefore it would seem absurd for us to deny it, because that is what we are.

JK: Yes, quite, quite. So it means really moving away—again this is only words—from everything that man has put together as a means of timelessness.

DB: Can we say that none of the methods that man uses outwardly is going to free the mind from time?

JK: Absolutely.

DB: Every method implies time.

JK: Of course. It is so simple.

DB: We start out immediately by setting up the whole structure of time; the whole notion of time is presupposed before we start.

JK: Yes, quite. But how will you convey this to another? How will you, me, or X, convey this to a man who is caught in time and will resist it, fight it, because he says there is no other way? How will you convey this to him?

DB: I think that you can only convey it to somebody who has gone into it; you are not likely to convey it at all to somebody you just pick up off the street!

JK: So then, what are we doing? As that cannot be conveyed through words, what is a man to do? Would you say that to resolve a problem as it arises you have to go into it immediately, because otherwise you may do the most foolish thing and delude yourself that you have resolved it? Suppose I have a problem, any psychological problem. Can the mind realize, resolve it immediately? Not deceive itself, not resist it—you understand? But face it and end it.

DB: Well, with a psychological problem, that is the only way. Otherwise we would be caught in the very source of the problem.

JK: Of course. Would that activity end time, the psychological time that we are talking about?

DB: Yes, if we could bring this immediate action to bear on the problem, which is the self.

JK: One is greedy or envious. To end immediately greed, attachment, and so on, will that not give a clue to the ending of time?

DB: Yes, because any action which is not immediate has already brought in time.

JK: Yes, yes. I know that.

DB: The ending of time is immediate, right?

JK: Immediate, of course. Would that point out the wrong turn that mankind has taken?

DB: Yes, if man feels something is out of order psychologically, he then brings in the notion of time and the thought of becoming, and that creates endless problems.

JK: Would that open the door to this sense of time having no place inwardly? Which means, doesn't it, that thought has no place except outwardly?

DB: You are saying that thought is a process which is involved in time.

JK: Wouldn't you say that thought is the process of time? Because thought is based on experience, knowledge, memory, and response, which is the whole of time.

DB: Yes, but still we have often discussed a kind of thought that would be the response to intelligence. Let's try to put it that thought, as we have generally known it, is in time.

JK: Thought as we know it now is of time.

DB: Yes. I would agree, generally speaking.

JK: Generally speaking, thought is time.

DB: It is based on the notion of time.

JK: Yes, all right. But to me, thought itself is time.

DB: Thought itself creates time, right.

JK: Does it mean, when there is no time there is no thought?

DB: Well, no thought of that kind.

JK: No. There is no thought. I want just to go slowly, slowly.

DB. Could we say that there is a kind of thought which we have lived in which has been dominated by time?

JK: Yes, but that has come to an end.

DB: But there may be another kind of thought which is not dominated by time. . . . I mean, you were saying, you could still use thought to do some things.

JK: Of course, outwardly that's so.

DB: We have to be careful not to say that thought is necessarily dominated by time.

JK: Yes. I have to go from here to there, to my house; that needs time, thought, but I am not talking of that kind of time.

DB: So let's make it clear that you are talking of thought which is aimed at the mind, whose content is of the order of the mind.

JK: Yes. Would you say knowledge is time?

DB: Well, yes . . .

JK: All knowledge is time.

DB: Yes, in that it has been known, and may project into the future, and so on.

JK: Of course, the future, the past. Knowledge—science, mathematics, whatever it is—is acquired through time. I read philosophy, I read this or that, and the whole movement of knowledge involves time. See what I mean?

DB: I think we are saying that man has taken a wrong turn and got caught in this kind of knowledge which is dominated by time because it has become psychological knowledge.

JK: Yes. So he lives in time.

DB: He lives in time because he has attempted to produce knowledge of the nature of the mind. Are you saying that there is no real knowledge of the mind? Would you put it that way?

JK: The moment you use the word "knowledge," it implies time. When you end time, in the sense we are talking about, there is no knowledge as experience.

DB: We have to see what the word "experience" means.

JK: Experience, memory.

DB: People say, "I learn by experience, I go through something."

JK: Which is becoming!

DB: Well, let's get it clear. You see, there is a kind of experience, for example, in one's job, which becomes skill and perception.

JK: Of course, but that is quite different.

DB: We are saying there is no point in having experience of the mind, psychological experience.

JK: Yes, let's put it that way. Psychological experience is in time.

DB: Yes, and it has no point, because you cannot say, "As I become skilled in my job, I will become skilled in operating my mind, or skilled fundamentally."

JK: Yes. So where is this leading? I realize that knowledge is time; the brain realizes it, and sees the importance of time in a certain direction, and that there is no value in time at all in another direction. It is not a contradiction.

DB: I would put it that the value of time is limited to a certain direction or area, and beyond that, it has no value.

JK: Yes. So what is the mind or the brain without knowledge? You understand.

DB: Without psychological knowledge?

JK: Yes, I am talking psychologically.

DB: It is not so much that it is caught in time as that it is without psychological knowledge to organize itself.

JK: Yes.

DB: So we are saying that the brain feels it must organize itself by knowing psychologically all about itself.

JK: Is then the mind, the brain, disorder? Certainly not.

DB: No. But I think that people being faced with this might feel there would be disorder.

JK: Of course.

DB: I think what you are saying is that the notion of controlling yourself psychologically has no meaning.

JK: So knowledge of the "me"—the psychological knowledge—is time.

DB: Yes, I understand the totality of knowledge is "me," is time.

JK: So then what is existence without this? There is no time, there is no knowledge in the psychological sense, no sense of "me," then what is there? To come to that point, most people would say, "What a horror this is."

DB: Yes, because it seems there would be nothing.

JK: Nothing. You've led us to a blank wall.

DB: It would be rather dull [*laughs*]. It is either frightening or it is all right.

JK: But if one has come to that point, what is there? Would you say, because there is nothing, it is everything?

DB: Yes, I would accept that. I know that. That is true, it has all . . .

JK: No, it is nothing.

DB: No thing.

JK: No thing. That's right [*laughs*].

DB: A thing is limited, and this is not a thing because there are no limits. . . . At least, it has everything in potential.

JK: Wait, sir. If it is nothing, and so everything, so everything is energy.

DB: Yes. The ground of everything is energy.

JK: Of course. Everything is energy. And what is the source of this thing? Or is there no source of energy at all? Is there only energy?

DB: Energy just is. Energy is "what is." There is no need for a source. That is one approach, perhaps?

JK: No. If there is nothing, and therefore everything, and everything is energy . . . We must be very careful because here the Hindus have this idea too, which is that Brahman is everything. You understand? But that becomes an idea, a principle, and is then carried out. But the fact of it is there is nothing; therefore there is everything, and all that is cosmic energy. But what started this energy?

DB: Is that a meaningful question? We are not talking of time.

JK: I know we are not talking of time, but you see the Christians would say, "God is energy and he is the source of all energy." No?

DB: Well, the Christians have an idea of what they call the God-head, which is the very source of God too.

JK: And also the Hindus, the Arabic, and the Jewish worlds have this. Are we going against all that?

DB: It sounds similar in some ways.

JK: And yet not similar. We must be awfully careful.

DB: Many things like this have been said over the ages.

JK: Then is one just walking in emptiness? Is one living in emptiness?

DB: Well, that is not clear.

JK: There is nothing, and everything is energy. What is this [*points to his body*]?

DB: Well, this is a form within the energy.

JK: This, the body, is not different from energy. But the thing that is inside says, "I am totally different from that."

DB: The "I" encloses itself and says, "I am different. I am eternal."

JK: Now, wait a minute. Why has it done this? Why has the separation arisen? Is it because outwardly I identify with a house and so on, and that identification has moved inwardly?

DB: Yes. And the second point was that once we established a notion of something inward, then it became necessary to protect that. And therefore that built up the separation.

JK: Of course.

DB: The inward was obviously the most precious thing, and it would have to be protected with all our energy.

JK: Does it mean then that there is only the organism living—which is part of energy? There is no K, no "me" at all, except the passport, name, and form, otherwise nothing? And therefore there is everything, and therefore all is energy?

DB: Yes, the form has no independent existence.

JK: Yes. No, what I am saying is there is only the form. That's all.

DB: There is also the energy, you see.

JK: That is part of energy. So there is only this, the outward shell.

DB: There is the outward form in the energy.

JK: Do you realize what we have said, sir [*laughs*]? Is this the end of the journey?

DB: Well, no, I should think not.

JK: Has mankind journeyed through millennia to come to this? That I am nothing, and therefore I am everything, and all energy?

DB: Well, it can't be the end, in the sense that it might be the beginning.

JK: Wait. That is all I wanted you to begin with. The ending is the beginning, right? Now I want to go into that. You see, in the ending of all this—the ending of time, we will call it briefly—there is a new beginning. What is that? Because otherwise this seems so utterly futile. I am all energy and just the shell exists, and time has ended. It seems so futile.

DB: Yes, if we stop there . . .

JK: That's all.

DB: I think that really this is clearing the ground of all the debris, of all the confusion.

JK: Yes. So the ending is a beginning. But what is that? Beginning implies time also.

DB: Not necessarily. I think we said there could be a movement which had no time.

JK: That is why I want to make it clear.

DB: Yes, but it is hard to express. It is not a question of being static, but in some sense the movement has not the order of time. I think we would have to say that now.

JK: Yes. So we will use the word "beginning" and deprive it of time.

DB: Because ending and beginning are no special time. In fact they can be any time or no time.

JK: No time. Then what takes place? What is happening? Not to me, not to my brain. What is happening? We have said that when one denies time, there is nothing. After this long talk, nothing means everything. Everything is energy. And we have stopped there. But that isn't the end.

DB: No.

JK: That is not the end. Then what is going on? Is that creation?

DB: Yes, something like that.

JK: But not the art of creating like writing or painting.

DB: Perhaps later we can discuss what we mean by creating.

Two

Cleansing the Mind of the Accumulation of Time

2 APRIL 1980, OJAI, CALIFORNIA

JIDDU KRISHNAMURTI: We were saying that psychological time is conflict, that time is the enemy of man. And that enemy has existed from the beginning of man. And we asked why has man from the beginning taken a "wrong turn," a "wrong path." And, if so, is it possible to turn man in another direction in which he can live without conflict? Because, as we said yesterday, the outer movement is also the same as the inner movement. There is no separation between inner and outer. It is the same movement. And we asked whether we were concerned deeply and passionately to turn man in another direction so that he doesn't live in time, with a knowledge only of the outer things. The religions, the politicians, the educators have failed: They have never been concerned about this. Would you agree to that?

DAVID BOHM: Yes. I think the religions have tried to discuss the eternal values beyond time but they don't seem to have succeeded.

JK: That is what I want to get at. To them it has been an idea, an ideal, a principle, a value, but not an actuality, and most of the religious people have their anchor in a belief, in a principle, in an image, in knowledge, in Jesus, or in something or other.

DB: Yes, but if you were to consider all the religions, say the various forms of Buddhism, they try to say this very thing which you are saying, to some extent.

JK: To some extent, but what I am trying to get at is why has man never confronted this problem? Why haven't we said, "Let's end conflict"? Instead we have been encouraged, because through conflict we think there is progress.

DB: It can be a certain source of stimulus to try to overcome opposition.

JK: Yes, sir, but if you and I see the truth of this, not in abstraction but actually, deeply, can we act in such a way that every issue is resolved instantly, immediately, so that psychological time is abolished? And as we asked yesterday, when you come to that point where there is nothing and there is everything, where all that is energy, when time ends, is there a beginning of something totally new? Is there a beginning which is not enmeshed in time? Now how shall we discover it? Words are necessary to communicate. But the word is not that thing. So what is there when all time ends? Psychological time, not time of . . .

DB: . . . the time of day.

JK: Yes. Time as the "me," the ego, and when that completely comes to an end, what is there that begins? Could we say that out of the ashes of time there is a new growth? What is that which begins . . . No, that word "begins" implies time too.

DB: Whatever we mean, that which arises.

JK: That which arises. What is it?

DB: Well, as we said yesterday, essentially it is creation, the possibility of creation.

JK: Yes, creation. Is that it? Is something new being born?

DB: It is not the process of becoming.

JK: Oh, no, that is finished. Becoming is the worst—that is time, that is the real root of this conflict. We are trying to find out what happens when the "I," which is time, has completely come to an end. I believe the Buddha is supposed to have said "nirvana." And the Hindus call it "moksha." I don't know whether the Christians call it "heaven" . . .

DB: The Christian mystics have had some similar state . . .

JK: Similar, yes. But you see, the Christian mystics, as far as I understand it, are rooted in Jesus, in the Church, in the whole belief. They have never gone beyond it.

DB: Yes, well, that seems so. As far as I know anyway.

JK: Now, we have said belief, attachment to all that, is out, finished. That is all part of the "I." Now, when there is that absolute cleansing of the mind from the accumulation of time, which is the essence of the "me," what takes place? Why should we ask what takes place?

DB: You mean it is not a good question?

JK: I am just asking myself why should we ask that? Is there behind it a subtle form of hope? A subtle form of saying I have reached that point, there is nothing? Then that's a wrong question. Wouldn't you consider that so?

DB: Well, it invites you to look for some hopeful outcome.

JK: If all endeavour is to find something beyond the "me," the endeavour and the thing that I may find are still within the orbit of "me." So I have no hope. There is no sense of hope. There is no sense of wanting to find anything.

DB: What is then moving you to enquire?

JK: My enquiry has been to end conflict.

DB: Yes, we have then to be careful. We are liable to produce a hope of ending conflict.

JK: No, no. There is no hope. I end it. The moment I introduce the word "hope" there is a feeling of the future.

DB: Yes, that is desire.

JK: Desire—and therefore it is of time. So I—the mind—puts all that aside completely; I mean it, completely. Then what is the essence of all this? Is my mind still seeking or groping after something intangible that it can capture and hold? If that is so, it is still part of time.

DB: Well, that is still desire.

JK: Desire and a subtle form of vanity.

DB: Why vanity?

JK: Vanity in the sense "I have reached."

DB: Self-deception.

JK: Deception and all forms of illusion arise from that. So it is not that. I am clearing the decks as we go along.

DB: Essentially it seems that you are clearing the movement of desire in its subtle forms.

JK: In its subtle forms. So desire too has been put away. Then there is only mind, right?

DB: Yes, but then we have to ask what is meant by nature, if all is mind, because nature seems somewhat independent.

JK: But we have also said that all the universe is the mind.

DB: You mean to say nature is the mind?

JK: Part of the mind.

DB: The universal mind?

JK: Yes.

DB: Not a particular mind?

JK: The particular mind then is separate, but we are talking of mind.

DB: You see, we have to make it clear, because you are saying that nature is the creation of universal mind, though nevertheless nature has a certain reality.

JK: That is all understood.

DB: But it is almost as if nature were the thought of the universal mind.

JK: It is part of it. I am trying to grope towards the particular and coming to an end; then there is only the mind, the universal mind, right?

DB: Yes. We have been discussing the particular mind groping through desire, and we said if all of that stopped . . .

JK: That is just my point. If all that has completely come to an end, what is the next step? *Is* there any next? We said yesterday there is a beginning, but that word implies part of time.

DB: We won't say so much beginning, perhaps ending.

JK: The ending—we have said that.

DB: But now is there something new?

JK: Is there something which the mind cannot capture?

DB: Which mind, the particular or the universal?

JK: The particular has ended.

DB: Yes. You are saying the universal mind cannot capture it either?

JK: That is what we are finding out.

DB: Are you saying there is a reality—or something—beyond universal mind?

JK: Are we playing a game of peeling off one thing after another? Like an onion skin, and at the end there is only tears and nothing else?

DB: Well, I don't know.

JK: Because we said there is the ending, then the cosmic, the universal mind, and, beyond, is there something more?

DB: Well, would you say this "more" is energy? That energy is beyond the universal mind?

JK: I would say yes, because the universal mind is part of that energy.

DB: That is understandable. In a way the energy is alive, you are saying?

JK: Yes, yes.

DB: And also intelligent?

JK: Wait a minute.

DB: In some way . . . insofar as it is mind.

JK: Now, if that energy is intelligent, why has it allowed man to move away in the wrong direction?

DB: I think that that may be part of a process, something that is inevitable in the nature of thought. You see, if thought is going to develop, that possibility must exist. To bring about thought in man . . .

JK: Is that the original freedom for man? To choose?

DB: No. That is, thought has to have the capacity to make this mistake.

JK: But if that intelligence was operating, why did it allow this mistake?

DB: Well, we can suggest that there is a universal order, a law.

JK: All right. The universe functions in order.

DB: Yes, and it is part of the order of the universe that this particular mechanism can go wrong. If a machine breaks down, it is not disorder in the universe; it is merely part of the universal order.

JK: Yes. In the universal order there is disorder, where man is concerned.

DB: It is not disorder at the level of the universe.

JK: No. At a much lower level.

DB: At the level of man it is disorder.

JK: And why has man lived from the beginning in this disorder?

DB: Because he is still ignorant, he still hasn't seen the point.

JK: But he is part of the whole, and in one tiny corner man exists and has lived in disorder. And this enormous intelligence has not . . .

DB: Yes, you could say that the possibility of creation is also the possibility of disorder; that if man had the possibility of being creative, there would also be the possibility of a mistake. He could not be fixed like a machine, always to operate in perfect order. The intelligence would not have turned him into a machine that would be incapable of disorder.

JK: No, of course not. So is there something beyond the cosmic order? Mind?

DB: Are you saying that the universe, that that mind has created nature which has an order, which is not merely going around mechanically? It has some deeper meaning?

JK: That is what we are trying to find out.

DB: You are bringing in the whole universe as well as mankind. What makes you do this? What is the source of this perception?

JK: Let's begin again. There is the ending of the "me" as time, and so there is no hope; all that is finished, ended. In the ending of it, there is that sense of nothingness, which is so. And nothingness is this whole universe.

DB: Yes, the universal mind, the universal matter.

JK: The whole universe.

DB: I am just asking: What led you to say that?

JK: Ah, I know. To put it very simply, division has come to an end. Right? The division created by time, created by thought, created by this education, and so on—all that. Because it has ended, the other is obvious.

DB: You mean that without the division then the other is there—to be perceived?

JK: Not to be perceived; it is there.

DB: But then how does one come to be aware that it is there?

JK: I don't think one becomes aware of it.

DB: Then what leads you to say it?

JK: Would you say it is? Not I perceive it, or it is perceived.

DB: Yes. It is.

JK: It is.

DB: You could almost say that it is saying it. In some sense, you seem to be suggesting that it is what is saying.

JK: Yes. I didn't want to put it—I am glad you put it like that! Where are we now?

DB: We are saying that the universe is alive, as it were, it is mind, and we are part of it.

JK: We can only say we are part of it when there is no "I."

DB: No division.

JK: No division. I would like to push it a little further; is there something beyond all this?

DB: Beyond the energy, you mean?

JK: Yes. We said nothingness, that nothingness is everything, and so it is that which is total energy. It is undiluted, pure, uncorrupted energy. Is there something beyond that? Why do we ask it?

DB: I don't know.

JK: I feel we haven't touched it; I feel there is something beyond.

DB: Could we say this something beyond is the ground of the whole? You are saying that all this emerges from a more inward ground?

JK: Yes, there is another—I must be awfully careful here. You know one must be awfully careful not to be romantic, not to have illusions, not to have desire, not even to search. It must happen. You follow what I mean?

DB: We are saying the thing must come from that. Whatever you are saying must come from that.

JK: From that. That's it. It sounds rather presumptuous!

DB: Without your actually seeing it. It is not that you look at it and say that is what I have seen.

JK: Oh, no. Then it is wrong.

DB: There is the division already. Of course, it is easy to fall into delusion with this sort of thing.

JK: Yes, but we said delusion exists as long as there is desire and thought. That is simple. And desire and thought are part of the "I," which is time. When desire and time are completely ended, then there is absolutely nothing, and therefore that is the universe, that emptiness which is full of energy. We can put a stop there . . .

DB: Because we haven't yet seen the necessity for going beyond the energy. We have to see that as necessary.

JK: I think it is necessary.

DB: Yes, but it has to be seen. We have to bring out why it is necessary.

JK: Why is it necessary? Tentatively, because there is something else that is operating, there is something much more—much—I don't know how to put it—much greater. I am going slowly, slowly. What I am trying to say is I think there is something beyond that. When I say, "I think," you know what I mean.

DB: I understand, yes.

JK: There is something beyond that. How can we talk about it? You see, energy exists only when there is emptiness. They go together.

DB: This pure energy you talk about is emptiness. Are you suggesting there is that which is beyond the emptiness, the ground of the emptiness?

JK: Yes.

DB: Would that be something in the nature of a substance? You see the question is if it is not emptiness, then what is it?

JK: I don't quite follow your question.

DB: Well, you say something beyond emptiness, other than emptiness. I think we can follow to the energy and the emptiness. Now, if we suggest something other to that, to the emptiness . . .

JK: Oh, yes, there is something other.

DB: Yes, then that other must be different from the emptiness. Something other to emptiness, which therefore is not emptiness. Does that make sense?

JK: Then it is substance.

DB: Yes, that is what is implied: If it is not emptiness, it is substance.

JK: Substance is matter, is it not?

DB: Not necessarily, but having the quality of substance.

JK: What do you mean by that?

DB: Matter is a form of substance in the sense that it is energy, but having the form of substance as well, because it has a constant form and it resists change. It is stable; it maintains itself.

JK: Yes. But when you use the word "substance," meaning beyond emptiness, does that word convey that meaning?

DB: Well, we are exploring the possible meaning of what you want to say. If you are saying it is not emptiness, then it would not be substance as we know it in matter. But we can see a certain quality which belongs to substance in general; if it has that quality, we could use the word "substance," extend the meaning of it.

JK: I understand. So could we use the word "quality"?

DB: The word "quality" is not necessarily the emptiness; energy could have the quality of emptiness, you see. And therefore it is something else. Something other might have the quality of substance. That is the way I see it. And is that what you are trying to say?

JK: There is something beyond emptiness. How shall we tackle it?

DB: Firstly, what leads you to say this?

JK: Simply the fact that there is. We have been fairly logical all along, we have not been caught in any illusions so far. And can we keep that same kind of watchfulness, in which there is no illusion, to find out—not find out—for that which is beyond emptiness to come down to earth? Come down to earth in the sense to be communicated. You follow what I mean?

DB: Yes. Well, we could come back to the question before: Why hasn't it come down?

JK: Why hasn't it come down? Has man been ever free from the "I"?

DB: No. Not generally speaking.

JK: No. And it demands that the "I" ends.

DB: I think we could look at it this way: that the ego becomes an illusion of that substance. You feel the ego is a substance too in some way.

JK: Yes, the ego is substance.

DB: And therefore that substance seems to be . . .

JK: . . . untouchable.

DB: But that ego is an illusion of the true substance—it may be that the mind tries to create some sort of illusion of that substance.

JK: That is an illusion. Why do you relate it to the other?

DB: In the sense that if the mind thinks it already has this substance, then it will not be open . . .

JK: Of course not. Can that thing ever be put into words? It is not a question of avoiding something or trying to slither out of some conclusion. But you see, so far we have put everything into words.

DB: Well, I think that once something is properly perceived, then after a while the words come to communicate it.

JK: Yes, but can that be perceived? And therefore be communicable?

DB: This thing beyond, would you also say it is alive? Life beyond emptiness, is that still life? Living?

JK: Living, yes. Oh, yes.

DB: And intelligent?

JK: I don't want to use those words.

DB: They are too limited?

JK: "Living," "intelligence," "love," "compassion"—they are all too limited. You and I are sitting here. We have come to a point, and there is that thing which perhaps later on might be put into words without any sense of pressure, and so without any illusion. Don't you feel, not feel, don't you see beyond the wall? You know what I mean? We have come to a certain point, and we are saying there is something still more. You understand? There is something behind all that. Is it palpable? Can we touch it? Is it something that the mind can capture? You follow?

DB: Yes. Are you saying it is not?

JK: I don't think it is possible for the mind to capture it . . .

DB: Or grasp it?

JK: Grasp it, understand . . . for the mind even to look at it. You are a scientist, you have examined the atom, and so on. Don't you, when you have examined all that, feel there is something much more, beyond all that?

DB: You can always feel that there is more beyond that, but it doesn't tell you what it is. It is clear that whatever one knows is limited.

JK: Yes.

DB: And there must be more beyond.

JK: How can that communicate with you, so that you, with your scientific knowledge, with your brain capacity, can grasp it?

DB: Are you saying it can't be grasped?

JK: No. How can you grasp it? I don't say you can't grasp it. Can you grasp it?

DB: Look, it is not clear. You were saying before that it is ungraspable by . . .

JK: Grasp in the sense, can your mind go beyond theories? What I am trying to say is can you move into it? Not move in the sense of time and all that. Can you enter it? No, those are all words. What is beyond emptiness? Is it silence?

DB: Isn't that similar to emptiness?

JK: Yes, that is what I am getting at. Move step by step. Is it silence? Or is silence part of emptiness?

DB: Yes, I should say that.

JK: I should say that too. If it is not silence, could we—I am just asking—could we say it is something absolute? You understand?

DB: Well, we could consider the absolute. It would have to be something totally independent; that is what "absolute" really means. It doesn't depend on anything.

JK: Yes. You are getting somewhere near it.

DB: Entirely self-moving, as it were, self-active.

JK: Yes. Would you say everything has a cause, and that has no cause at all?

DB: You see, this notion is already an old one. This notion has been developed by Aristotle, that this absolute is the cause of itself.

JK: Yes.

DB: It has no cause, in a sense. That is the same thing.

JK: You see, the moment you said Aristotle . . . It is not that. How shall we get at this? Emptiness is energy, and that emptiness exists in silence, or the other way round. It doesn't matter, right? Oh, yes, there is something beyond all this. Probably it can never be put into words. But it must be put into words. You follow?

DB: You are saying that the absolute must be put into words, but we feel it can't be? Any attempt to put it into words makes it relative.

JK: Yes. I don't know how to put all this.

DB: I think that we have a long history of danger with the absolute. People have put it in words, and it has become very oppressive.

JK: Leave all that. You see, being ignorant of what other people have said, Aristotle and the Buddha and so on, has an advantage. You understand what I mean? An advantage in the sense that the mind is not coloured by other people's ideas, not caught in other people's statements. All that is part of our conditioning. Now, to go beyond all that! What are we trying to do?

DB: I think to communicate regarding this absolute, this beyond.

JK: I took away that word "absolute" immediately.

DB: Then whatever it is, the beyond emptiness and silence.

JK: Beyond all that. There is beyond all that. All that is something part of an immensity.

DB: Yes, well even the emptiness and silence is an immensity, isn't it? The energy is itself an immensity.

JK: Yes, I understand that. But there is something much more immense than that. Emptiness and silence and energy are immense, really immeasurable. But there is something that is—I am using the word "greater"—than that. Why do you accept all this?

DB: I am just considering. I am looking at it. One can see that whatever you say about emptiness, or about any other thing, there is something beyond.

JK: No, as a scientist, why do you accept—not accept, forgive me for using that word—why do you even move along with this?

DB: Because we have come this far step by step, seeing the necessity of each step.

JK: You see all that is very logical, reasonable, sane.

DB: And also one can see that it is so, right?

JK: Yes. So if I say there is something greater than all this silence, energy, would you accept that? Accept in the sense that up to now we have been logical.

DB: We will say that whatever you speak of there is certainly something beyond it. Silence, energy, whatever, then there is always room logically for something beyond that. But the point is this: that even if you were to say there is something beyond that, still you logically leave room for going again beyond that.

JK: No.

DB: Well, why is that? You see, whatever you say, there is always room for something beyond.

JK: There is nothing beyond.

DB: Well, that point is not clear, you see.

JK: There is nothing beyond it. I stick to that. Not dogmatically or obstinately. I feel that is the beginning and the ending of everything. The ending and the beginning are the same, right?

DB: In which sense? In the sense that you are using the beginning of everything as the ending?

JK: Yes. Right? You would say that?

DB: Yes. If we take the ground from which it comes, it must be the ground to which it falls.

JK: That's right. That is the ground upon which everything exists, space . . .

DB: . . . energy . . .

JK: . . . energy, emptiness, silence, all that is. All that. Not ground, you understand?

DB: No, it is just a metaphor.

JK: There is nothing beyond it. No cause. If you have a cause, then you have ground.

DB: You have another ground.

JK: No. That is the beginning and the ending.

DB: It is becoming more clear.

JK: That's right. Does that convey anything to you?

DB: Yes. Well, I think that it conveys something.

JK: Something. Would you say further there is no beginning and no ending?

DB: Yes. It comes from the ground, goes to the ground, but it does not begin or end.

JK: Yes. There is no beginning and no ending. The implications are enormous. Is that death? Not death in the sense I will die, but the complete ending of everything?

DB: You see, at first you said that the emptiness is the ending of everything, so in what sense is this more, now? Emptiness is the ending of things, isn't it?

JK: Yes, yes. Is that death, this emptiness? Death of everything the mind has cultivated. This emptiness is not the product of the mind, of the particular mind.

DB: No, it is the universal mind.

JK: That emptiness is that.

DB: Yes.

JK: That emptiness can only exist when there is death—total death—of the particular.

DB: Yes.

JK: I don't know if I am conveying this.

DB: Yes, that is the emptiness. But then you are saying that, in this ground, death goes further?

JK: Oh, yes.

DB: So we are saying the ending of the particular, the death of the particular, is the emptiness, which is the universal. Now, are you going to say that the universal also dies?

JK: Yes, that is what I am trying to say.

DB: Into the ground.

JK: Does it convey anything?

DB: Possibly, yes.

JK: Just hold it a minute. Let's see it. I think it conveys something, doesn't it?

DB: Yes. Now, if the particular and the universal die, then that is death?

JK: Yes. After all, an astronomer says everything in the universe is dying, exploding, dying.

DB: But of course you could suppose that there was something beyond.

JK: Yes, that is just it.

DB: I think we are moving. The universal and the particular. First the particular dies into the emptiness, and then comes the universal.

JK: And that dies too.

DB: Into the ground, right?

JK: Yes.

DB: So you could say the ground is neither born nor dies.

JK: That's right.

DB: Well, I think it becomes almost inexpressible if you say the universal is gone, because expression is the universal.

JK: You see, I am just explaining: Everything is dying, except that. Does this convey anything?

DB: Yes. Well, it is out of that that everything arises, and into which it dies.

JK: So that has no beginning and no ending.

DB: What would it mean to talk of the ending of the universal? What would it mean to have the ending of the universal?

JK: Nothing. Why should it have a meaning if it is happening? What has that to do with man? You follow what I mean? Man, who is going through a terrible time. What has that got to do with man?

DB: Let's say that man feels he must have some contact with the ultimate ground in his life, otherwise there is no meaning.

JK: But it hasn't. That ground hasn't any relationship with man. He is killing himself; he is doing everything contrary to the ground.

DB: Yes, that is why life has no meaning for man.

JK: I am an ordinary man; I say, all right, you have talked marvellously, but what has that got to do with me? Will that or your talk help me to get over my ugliness? My quarrels with my wife or whatever it is?

DB: I think I would go back, and say we went into this logically starting from the suffering of mankind, showing it originates in a wrong turning, that leads inevitably . . .

JK: Yes, but man asks, help me to get past the wrong turn. Put me on the right path. And to that one says please don't *become* anything.

DB: Right. What is the problem then?

JK: He won't even listen.

DB: Then it seems to me that it is necessary for the one who sees this to find out what is the barrier to listening.

JK: Obviously you can see what is the barrier.

DB: What is the barrier?

JK: "I."

DB: Yes, but I meant more deeply.

JK: More deeply, all your thoughts, deep attachments—all that is in your way. If you can't leave these, then you will have no relationship with that. But man doesn't want to leave these.

DB: Yes, I understand. What he wants is the result of the way he is thinking.

JK: What he wants is some comfortable, easy way of living without any trouble, and he can't have that.

DB: No. Only by dropping all this.

JK: There must be a connection. There must be some relationship with the ground and this, some relationship with ordinary man. Otherwise, what is the meaning of living?

DB: That is what I was trying to say before. Without this relationship . . .

JK: . . . there is no meaning.

DB: And then people invent meaning.

JK: Of course.

DB: Even going back, the ancient religions have said similar things, that God is the ground, so they say seek God, you know.

JK: Ah, no, this isn't God.

DB: No, it is not God, but it is playing the same part. You could say that "God" is an attempt to put this notion a bit too personally perhaps.

JK: Yes. Give them hope, give them faith, you follow? Make life a little more comfortable to live.

DB: Well, are you asking at this point how is this to be conveyed to the ordinary man? Is that your question?

JK: More or less. And also it is important that he should listen to this. You are a scientist. You are good enough to listen because we are friends. But who will listen among your other friends? I feel that if one pursues this, we will have a marvellously ordered world.

DB: Yes. And what will we do in this world?

JK: Live.

DB: But, I mean, we said something about creativity . . .

JK: Yes. And then if you have no conflict, no "I"; there is something else operating.

DB: Yes, it is important to say that, because the Christian idea of heaven as perfection may seem rather boring because there is nothing to do!

JK: That reminds me of a good joke! You are waiting for the joke [*laughs*]? A man dies and goes to Saint Peter, and Saint Peter says, "You have lived a fairly good life, you have not cheated too much, but before you enter heaven I must tell you one thing. Here we are all bored. We are all awfully serious; God never laughs. And every angel is moody, depressed. Unless you want to enter this world, hesitate. So before you come in perhaps you would like to go down below and see what it is like. It's up to you. Ring that bell, the lift will come up. You get into it and go down." So the chap rings the bell, goes down, and the gates open. And he is met by the most beautiful girls. And he says, "By Jove, this is the life. May I go up and tell Peter?" And he gets into the lift, goes up and says, "Sir, it was very good of you to offer me the choice. I prefer it down below." And Peter says, "I thought so!" So the man rings the bell and goes down again. The gate opens and two people meet him and beat him up, push him around, and so on. He protests, "Just a minute. A minute ago I came here, you treated me like a king!" "Ah, then you were a tourist [*laughs*]!" Sorry. From the sublime to the ridiculous, which is good too [*laughs*].

We must continue this discussion some other time, because it is something that has to be put into orbit.

D B: It seems impossible.

J K: We'll see. We have gone pretty far.

Three

Why Has Man Given Supreme Importance to Thought?

8 APRIL 1980, OJAI, CALIFORNIA

JIDDU KRISHNAMURTI: What shall we talk about?

DAVID BOHM: One point relating to what we discussed before; I was reading somewhere that a leading physicist said that the more we understand the universe, the more pointless it seems, the less meaning it has. And it occurred to me that in science there may be an attempt to make the material universe the ground of our existence, so that it may have meaning physically but not . . .

JK: . . . any other meaning. Quite.

DB: And the question that we might discuss is this ground which we were talking about the other day. Is it indifferent to mankind, as the physical universe appears to be?

JK: Good question. Let's get it clear.

DB: Not only physicists but geneticists, biologists, have tried to reduce everything to the behaviour of matter—atoms, genes, DNA, and so on. And the more they study it, then the more they feel it has no meaning, it is just going on. Though it has meaning physically, in the sense that we can understand it scientifically, it has no deeper meaning than that.

JK: I understand that.

DB: And, of course, perhaps that notion has penetrated because in the past people were more religious and felt that the ground of our existence was in something beyond matter—God or whatever they wished to call it. And that gave them a sense of deep meaning to the whole of their existence, which has now gone away. That is one of the difficulties of modern life, the sense that it doesn't mean anything.

JK: So have the religious people invented something which has a meaning?

DB: They may well have done so. You see, feeling that life has no meaning, they may have invented something beyond the ordinary. Something which is eternal . . .

JK: . . . timeless, nameless.

DB: . . . and independent—the absolute, they call it.

JK: Seeing that the way we live, genetically and all the rest of it, has no meaning, some clever erudite people have said, "We will give it a meaning."

DB: Well, I think it happened before that. In the past people somehow gave meaning to life, long before science had been very much developed, in the form of religion. And science came along and began to deny this religion.

JK: Quite. I understand that.

DB: And people no longer believe in the religious meaning. Perhaps they never were able to believe in it entirely anyway.

JK: So how does one find out if life has a meaning beyond this? How does one find out? They have tried meditation; they have tried every form of self-torture, isolation, becoming a monk, a sannyasi, and so on. But they may also be deceiving themselves thoroughly.

DB: Yes. And that is in fact why the scientists have denied it all, because the story told by the religious people is not plausible, you see.

JK: Quite. So how does one find out if there is something more than the merely physical? How would one set about it?

DB: Well, we have been discussing in the past few days the notion of some ground which is beyond matter, beyond the emptiness.

JK: But suppose you say it is so, and I say that is another illusion.

DB: The first point is perhaps we could clear this up: You see, if this ground is indifferent to human beings, then it would be the same as the scientists' ground in matter.

JK: Yes. What is the question?

DB: Is the ground indifferent to mankind? You see, the universe appears to be totally indifferent to mankind. It is immense vastness; it pays no attention to us; it may produce earthquakes and catastrophes; it might wipe us out; it is essentially not interested in mankind.

JK: I see what you mean, yes.

DB: It does not care whether man survives or does not survive, if you want to put it that way.

JK: Right. I understand the question.

DB: Now, I think that people felt that God was a ground who was not indifferent to mankind. You see, they may have invented it, but that is what they believed. And that is what gave them possibly . . .

JK: . . . tremendous energy. Quite.

DB: Now, I think the point is would *this* ground be indifferent to mankind?

JK: How would you find out? What is the relationship of this ground to man, and man's relationship to it?

DB: Yes, that is the question. Does man have some significance to it? And does it have significance to man? May I add one more point? I was discussing with somebody who was familiar with the Middle East and traditions of mysticism; he told me that in these traditions they not only say that what we call this ground, this infinite, has some significance, but that what man does has ultimately some significance.

JK: Quite, quite. Suppose one says it has—otherwise life has no meaning, nothing has any meaning. How would one—not prove—how would one find out? Suppose you say this ground exists, as I said the other day. Then the next question is what relationship has that to man? And man to it? How would one discover or find out or touch it—if the ground exists at all? If it doesn't exist, then really man has no meaning at all. I mean, I die and you die and we all die, and what is the point of being virtuous, what is the point of being happy or unhappy, just carrying on? How would you show that the ground exists? In scientific terms, as well as the feeling of it, the nonverbal communication of it?

DB: When you say scientific, do you mean rational?

JK: Yes, rational.

DB: So something that we can actually touch.

JK: Sense—better than touch—sense. Scientific: We mean by that rational, logical, sane, many can come to it.

DB: Yes, it is public.

JK: It isn't just one man's assertion. It would be scientific. I think it can be shown, but with all things one must *do* it, not just talk about it. Can I—or you—say the ground exists? The ground has certain demands, which are there must be absolute silence, absolute emptiness, which means no sense of egotism in any form, right? Would you tell me that? Am I willing to let go all my egotism, because I want to prove it, I want to show it, I want to find out if what you are saying is actually true? So am I willing to say, "Look, complete eradication of the self"? Would all of us, ten of us, be willing to do that?

DB: I think I can say that perhaps in some sense one is willing, but there may be another sense in which the willingness is not subject to one's conscious effort or determination.

JK: No, wait. So we go through all that.

DB: We have to see that ...

JK: It is not will; it is not desire; it is not effort.

DB: Yes, but when you say willingness, it contains the word "will," for example.

JK: Willingness, in the sense, "Go through that door." Or, am I, are we, willing to go through that particular door to find that the ground exists? You ask me that. I say, agreed, I will. I will in the sense of not exercising will and all that. What are the facets or the qualities or the nature of the self? We go into that. You point it out to me and I say, "Right." Can we, ten of us, do it? Not be attached, not have fear—you follow?—the whole business of it. No belief, absolute, rational—you know—observation. I think if ten people do it, any scientist will accept it. But there are no ten people.

DB: I see. We have to have the thing done together publicly . . .

JK: . . . that's it . . .

DB: . . . so that it becomes a real fact.

JK: A real fact in the sense that people accept it. Not something based on illusion, belief, and all the rest of that.

DB: A fact—that which is actually done.

JK: Now, who will do this? The scientists say that the thing is all illusory, nonsense. But there are others who say, X who says, "It is not nonsense; there *is* a ground. And if you do these things, it will be there."

DB: Yes, but I think that some of the things you say may not in the beginning entirely make sense to the person you talk with.

JK: Yes, quite, because he isn't even willing to listen.

DB: But also his whole background is against it. You see, the background gives you the notion of what makes sense and what doesn't. Now, when you say, for example, one of the steps is not to bring in time . . .

JK: Ah, that's much more difficult.

DB: Yes, but it is fairly crucial.

JK: But wait. I wouldn't begin with time; I would begin at the schoolboy level [*laughs*].

DB: But you are going eventually to reach those more difficult points.

JK: Yes. But begin at the schoolboy level and say, "Look, do these things."

DB: Well, what are they? Let's go over them.

JK: No belief.

DB: A person may not be able to control what he believes; he may not even know what he believes.

JK: No, don't control anything. Observe that you have belief, you cling to the belief, belief gives you a sense of security, and so on. And that belief is an illusion; it has no reality.

DB: You see, I think if we were to talk to scientists like that, they might say they were not sure about it, because they believe in the existence of the material world.

JK: You don't *believe* the sun rises and sets. It is a fact.

DB: Yes, but the scientist believes. You see, there have been long arguments about this; there is no way to prove that it exists outside my mind, but I believe it anyway. This is one of the questions which arises. Scientists actually have beliefs. One will believe that this theory is right, and the other believes in a different one.

JK: No. I have no theories. I don't have any theories. I start at the schoolboy level by saying, "Look, don't accept theories, conclusions, don't cling to your prejudices." That is the starting point.

DB: Perhaps we had better say don't hold to your theories, because somebody might question you if you say you have no theories. They would immediately doubt that, you see.

JK: I have no theories. Why should I have theories?

QUESTIONER: If I am a scientist, I would also say I don't have theories. I don't see that the world which I construct for my scientific theories is also theoretical. I would call it fact.

JK: So we have to discuss what are facts. Right? I would say that facts are that which is happening, actually happening. Would you agree to that?

DB: Yes.

JK: Would the scientists agree to that?

DB: Yes. Well, I think that the scientists would say that what is happening is understood through the theories. You see, in science you do not understand what is happening except with the aid of instruments and theories.

JK: Now, wait, wait. What is happening out there? What is happening here?

DB: Let's go slowly. First, what is happening out there. The instruments and theories are needed even to . . .

JK: No.

DB: . . . have the facts about what is out there . . .

JK: What are the facts out there?

DB: You cannot find out without some kind of theory.

JK: The facts there are conflict. Why should I have a theory about it?

DB: I wasn't discussing that. I was discussing the facts about matter, you see, which the scientist is concerned with. He cannot establish those facts without a certain theory, because the theory organizes the facts for him.

JK: Yes, I understand that. That may be a fact. You may have theories about that.

DB: Yes. About gravitation, atoms—all those things depend on theories in order to produce the right facts.

JK: The right facts. So you start with a theory.

DB: A mixture of theory and fact. It is always a combination of theory and fact.

JK: All right. A combination of theory and fact.

DB: Now, if you say we are going to have an area where there isn't any such combination . . .

JK: That's it. Which is, psychologically. I have no theory about myself, about the universe, about my relationship with another. I have no theory. Why should I have? The only fact is mankind suffers, is miserable, confused, in conflict. That is a fact. Why should I have a theory about it?

DB: You must go slowly. You see, if you are intending to bring in the scientists, if this is to be scientific . . .

JK: . . . I will go very slowly . . .

DB: . . . so that we don't leave the scientists behind!

JK: Quite. Or leave me behind!

DB: Well, let's accept "part company," right? The scientists might say yes, psychology is the science with which we try to look inwardly, to investigate the mind. And they say various people—like Freud and Jung and others—have had theories. Now, we will have to make it clear why it has no point to make these theories.

JK: Because theory prevents the observation of what is actually taking place.

DB: Yes, but outside it seemed that the theory was helping that observation. Why the difference here?

JK: The difference? You can discover that; it is simple.

DB: Let's spell it out. Because if you want to bring in scientists, you must answer this question.

JK: We will answer it. What is the question?

DB: Why is it that theories are both necessary and useful in organizing facts about matter outwardly, and yet inwardly, psychologically, they are in the way, they are no use at all.

JK: Yes. What is theory? The meaning of the word "theory"?

DB: *Theōria* meant to see, to view, a kind of insight.

JK: To view? That's it. A way of looking.

DB: And the theory helps you to look at the outside matter.

JK: "Theory" means to observe.

DB: It is a way of observing.

JK: Can you observe psychologically what is going on?

DB: Let's say that when we look at matter outwardly, to a certain extent we do the observing.

JK: That is, the observer is different from the observed.

DB: Not only different, but their relationship is fixed, relatively at least, for some time.

JK: So we can move now, a little.

DB: This appears to be necessary in order to study matter. Matter does not change so fast, and it can be separated to some extent. We can then make a fairly constant way of looking. It changes but not immediately; it can be held constant for a while.

JK: Yes.

DB: And we call that theory.

JK: As you said, theory means a way of observing.

DB: It is the same as "theatre" in Greek.

JK: Theatre, yes, that's right. It is a way of looking. Now, where do we start? The common way of looking, the ordinary way of looking, the way of looking depending on the viewpoint of each person—the housewife, the husband? What do you mean by the way of looking?

DB: The same problem arose in the development of science. We began with what was called common sense, a common way of looking. Then scientists discovered that this was inadequate.

JK: They moved away from it.

DB: They moved away; they gave up some parts of it.

JK: That is what I am coming to. The common way of looking is full of prejudice.

DB: Yes, it is arbitrary and dependent on your background.

JK: Yes, all that. So can one be free of one's background, one's prejudice? I think one can.

DB: The question is whether a theory of psychology would be any help in doing this. The danger is that the theory itself might be a prejudice. If you tried to make a theory . . .

JK: That is what I am saying. That would become a prejudice.

DB: That would become a prejudice because we have nothing— we have not yet observed anything to found it on.

JK: So the common factor is that man suffers, right? That is the common factor. And the way of observing matters.

DB: Yes. I wonder whether scientists would accept that as the most fundamental factor of man.

JK: All right. Conflict?

DB: Well, they have argued about it.

JK: Take anything; it doesn't matter. Attachment, pleasure, fear.

DB: I think some people might object, saying we should take something more positive.

JK: Which is what?

DB: Simply, for example, some people might have said that rationality is a common factor.

JK: No, no, no! I won't call rationality a common factor. If people were rational, they wouldn't be fighting each other.

DB: We have to make this clear. Let's say in the past somebody like Aristotle might have said rationality is the common factor of man. Now your argument against it is that men are not generally rational.

JK: No, they are not.

DB: Though they might be, they are not. So you are saying that is not a fact.

JK: That's right.

Q: I think commonly scientists would say that there are many different human beings and that the common factor of mankind is that they are all striving for happiness.

JK: Is that the common factor? No. I won't accept that—that many human beings are striving for happiness.

Q: No. Human beings are all different.

JK: Agreed. Stay there.

Q: What I am saying is that this is the common theory, which people believe to be a fact.

JK: That is, each person thinks he is totally different from others.

Q: Yes. And they are all independently struggling for happiness.

JK: They are all seeking some kind of gratification. Would you agree to that?

DB: That is one common factor. But the reason I brought up rationality was that the very existence of science is based on the notion that rationality is common to man.

JK: But each person is seeking his own individuality.

DB: But, you see, science would be impossible if that were entirely true.

JK: Quite.

Q: Why?

DB: Because everybody would not be interested in the truth. The very possibility of scientific discovery depends on people feeling that this common goal of finding the truth is beyond personal satisfaction, because even if your theory is wrong you must accept that it is wrong, though it is not gratifying. That is, it becomes very disappointing for people, but they accept it, and say, well, that is wrong.

JK: I am not seeking gratification. I am a common man. You have brought up that scientists take for granted that human beings are rational.

DB: At least when they do science. They may agree that they are not very rational in private life, but they say that at least they are capable of being rational when they do scientific work. Otherwise it would be impossible to begin.

JK: So outwardly, in dealing with matter, they are all rational.

DB: At least they try to be, and they are to some extent.

JK: They try to be, but they become irrational in their relationships with other human beings.

DB: Yes. They cannot maintain it.

JK: So that is the common factor.

DB: Yes. It is important to bring out this point—that rationality is limited, and, as you say, the fundamental fact is that more generally they cannot be rational. They may succeed in some limited area.

JK: That's right. That is a fact.

DB: That is a fact, though we don't say it is inevitable or that it can't be changed.

JK: No. It is a fact.

DB: It is a fact that it has been, it has happened, it is happening.

JK: Yes. I, as a common human being, have been irrational. And my life has been totally contradictory, and so on, which is irrational. Now, can I as a human being change that?

DB: Let's see how we could proceed from the scientific approach. This would raise the question, why is everybody irrational?

JK: Because we have been conditioned that way. Our education, our religion, our everything.

DB: But that won't get us anywhere, because it leads to more questions: How did we get conditioned and so on.

JK: We can go into all that.

DB: But I meant that following that line is not going to answer.

JK: Quite. Why are we conditioned that way?

DB: For example, we were saying the other day that perhaps man took a wrong turning, established the wrong conditioning.

JK: The wrong conditioning right from the beginning. Or seeking security—security for myself, for my family, for my group, for my tribe—has brought about this division.

DB: Even then you have to ask why man sought this security in the wrong way. You see, if there had been any intelligence, it would have been clear that this whole thing has no meaning.

JK: Of course, you are going back to taking the wrong turn. How will you show me we have taken a wrong turning?

DB: Are you saying that we want to demonstrate this scientifically?

JK: Yes. I think the wrong turn was taken when thought became all important.

DB: What made it all important?

JK: Now, let's work it out. What made human beings enthrone thought as the only means of operation?

DB: Also, it would have to be made clear why, if thought is so important, it causes all the difficulties. These are the two questions.

JK: That is fairly simple. So thought has been made king, supreme. And that may be the wrong turn of human beings.

DB: You see, I think that thought became the equivalent of truth. People took thought to give truth, to give what is always true. There is the notion that we have knowledge—which may hold in certain cases for some time—but men generalize, because knowledge is always generalizing. When they got to the notion that it would always be so, this gave the thought of what is true. This gave thought supreme importance.

JK: You are asking, aren't you, why has man given thought such importance?

DB: I think he has slipped into it.

JK: Why?

DB: Because he did not see what he was doing. You see, in the beginning he did not see the danger . . .

Q: Just before, you said that the common ground for man is reason . . .

JK: Scientists say that.

Q: If you can show a person that something is true . . .

JK: You show it to me. It is true I am irrational. That is a fact; that is true.

Q: But for that you don't need reason. Observation is sufficient for that.

JK: No. One goes and fights; one talks about peace. One is irrational. Dr. Bohm is pointing out that scientists say man is rational, but the fact is that everyday life is irrational. Now we are asking, show us scientifically why it is irrational. That is, show man in what way he has slipped into this irrationality, why human beings have accepted this. We can say it is habit, tradition, religion. And the scientists also—they are very rational in their own field, but very irrational in their lives.

Q: And you suggested that making thought the king is the main irrationality?

JK: That is right. We have reached that point.

DB: But how did we slip into making thought so important?

JK: Why has man given importance to thought as the supreme thing? I think that is fairly easy. Because that is the only thing he knows.

DB: It doesn't follow that he would give it supreme importance.

JK: Because the things I know—the things thought has created, the images, all the rest of it—are more important than the things I don't know.

DB: But, you see, if intelligence were operating, he would not come to that conclusion. It is not rational to say that all that I know is all that is important.

JK: So man is irrational.

DB: He slipped into irrationality to say all that I know is all that is important. But why should man have done this?

JK: Would you say that the mistake is made because he clings to the known and objects to anything unknown?

DB: That is a fact, but it is not clear why he should.

JK: Because that is the only thing he has.

DB: But I am asking why he was not intelligent enough to see this.

JK: Because he is irrational.

DB: Well, we are going around in circles!

JK: I don't think so.

DB: Look, every one of these reasons you give is merely another example of man's irrationality.

JK: That is all I am saying. We are basically irrational, because we have given thought supreme importance.

Q: But isn't the step before that that thought has built up the idea that I exist?

JK: Ah, that comes a little later; we have to go step by step.

Q: Surely for the "me," the only thing that exists is thought.

JK: Would the scientists accept that?

DB: The scientist feels he is investigating the real nature of matter, independent of thought, ultimately independent anyway. He wants to know the way the universe is. He may be fooling himself, but he feels that it wouldn't be worth doing unless he believes he is finding an objective fact.

JK: So would you say that through the investigation of matter he is trying to find something, he is trying to find the ground?

DB: That's exactly it.

JK: But wait! Is that it?

DB: Precisely, yes.

JK: Now, the religious man says you can find it by becoming terribly rational in your life. He doesn't accept that he *is* rational but says he is irrational, in contradiction, and so on. So either he will have to clear up that first, step by step, or he can do the whole thing at one blow. Right? One accepts that one is irrational.

DB: But there is a difficulty. If you accept you are irrational, you stop, because you say, how can you begin?

JK: Yes. But if I accept I am irrational—wait a minute—completely, I am rational!

DB: You will have to make that more clear. You could say that man has been deluding himself into believing that he is already rational.

JK: I don't accept that.

DB: Now, if you don't accept this delusion, then you are saying that rationality will be there.

JK: No, I don't accept it. The *fact* is I am irrational and, to find the ground, I must become extremely rational in my life. That's all. Irrationality has been brought about by thought creating this idea of me as separate from everybody else. So can I, being irrational, find the cause of irrationality and wipe it out? If I can't do that, I cannot reach the ground which is the most rational. Would a scientist who is investigating matter accept that the ground exists at all?

DB: Well, tacitly he is assuming that it does.

JK: It does. Mr. X comes along and says it *does* exist. And you, the scientists, say, "Show it." Mr. X says, "I will show it to you. First become rational in your life. As a scientist you meet with other scientists, experimenting and being rational in that area, although irrational in your own life. First become rational in your life; begin here, rather than there." What would you say to all that? This must be done without effort, without desire, without will, without any sense of persuasion, otherwise you are back in the game.

DB: Well, let's try to put it like this: Even in science you could not pursue the science fully unless you were rational.

JK: Somewhat rational.

DB: Somewhat rational, but, eventually, the failure of rationality blocks science anyway. Scientists cling to their theories, and they become jealous and so on.

JK: That's it. That is all. The irrationality overcomes them.

DB: So then you could say you might as well look at the source of the whole irrationality.

JK: That is what I am saying.

DB: But now you have to make it clear that it really can be done.

JK: Oh, yes, I am showing it to you. I say, first recognize, see, observe, be aware that you are totally irrational.

DB: The word "totally" will cause trouble, because if you were totally irrational you couldn't even begin to talk.

JK: No, that is my question. I say one is totally irrational. First recognize it. Watch it. The moment you admit there is some part of me that is rational, who wants to wipe away the irrationality . . .

DB: It is not that, but there must be sufficient rationality to understand what you are talking about.

JK: Yes, of course.

DB: Essentially, I would rather put it that one is dominated by one's irrationality, even though there is enough rationality to discuss the question.

JK: I question that.

DB: You see, otherwise we couldn't begin to talk.

JK: But listen. We begin to talk. A few of us begin to talk because we are willing to listen to each other. We are willing to say we'll set aside any conclusions we have; we are willing to listen to each other.

DB: That is part of rationality.

JK: With some of us perhaps, but the vast majority is not willing to listen to us. Because we are concerned, serious enough to find out if the ground exists, that gives us rationality to listen to each other.

DB: Listening is necessary for rationality.

JK: Of course. Are we saying the same thing?

DB: Yes.

JK: The scientist, through the investigation of matter, hopes to reach the ground. We and X and Y say, "Let us become rational in our life." Which means that you and I and X and Y are willing to listen to each other. That's all. The very listening is the beginning of rationality. Some people won't listen to us or to anybody. So can we, who are listening, be somewhat rational and begin? That is all my point. This is being terribly logical, isn't it? So can we proceed from there?

Why has man brought about this irrationality in his life? A few of us can apparently throw off some part of irrationality, become somewhat rational and say, "Now, let's start. Let us start to find out why man lives this way. Now, what is the common dominant factor, common current, in all our lives." Obviously it is thought.

DB: Yes, that is so. Of course many people might deny that and say it is feeling or that something else is the major factor.

JK: Many people might say that, but thought is part of feeling.

DB: Yes, but that is not commonly understood.

JK: We will explain it. Feeling—if there was no thought behind it, would you be able to recognize it?

DB: Yes, I think this is a major difficulty in communication with some people.

JK: So we begin. There may be some who don't see this, but I want the three, X and Y and Z, to see it, because they have become somewhat rational, therefore they are listening to each other. They can say thought is the main source of this current.

DB: Then we have to say, what is thought?

JK: I think that is fairly simple. Thought brings about irrationality.

DB: Yes, but what is it? How do you know you are thinking? What do you mean by thinking?

JK: Thinking is the movement of memory, which is experience, knowledge, stored in the brain.

DB: Suppose we want to have rationality, which includes rational thought. Is rational thought only memory?

JK: Wait a minute. Let's be careful. If we are completely rational, there is total insight. That insight uses thought, and then it is rational.

DB: Then thought is not only memory?

JK: No, no.

DB: Well, I mean since it is being used by insight . . .

JK: No, insight uses thought.

DB: Yes, but what thought does is not just due to memory now.

JK: Wait a minute.

DB: Ordinarily thought runs on its own, it runs like a machine on its own, and it is not rational.

JK: Quite right.

DB: But when thought is the instrument of insight . . .

JK: Then thought is not memory.

DB: It is not based on memory.

JK: No, not based on memory.

DB: Memory is used, but it is not based on memory.

JK: Then what? Thought being limited, divisive, incomplete, can never be rational . . .

DB: Without insight.

JK: That's right. Now, how are we to have insight which is total rationality? Not the rationality of thought.

DB: I should call it rationality of perception.

JK: Yes, rationality of perception.

DB: Then thought becomes the instrument of that, so it has the same order.

JK: Now, how am I to have that insight? That is the next question. Isn't it? What am I to do, or not to do, to have this instant insight, which is not of time, which is not of memory, which has no cause, which is not based on reward or punishment? It is free of all that. Now, how does the mind have this insight? When I say "I" have the insight, that is wrong. Obviously. So how is it possible for a mind, which has been irrational and has become somewhat rational, to have that insight? It is possible to have that insight if your mind is free from time.

DB: Right. Let's go slowly because, you see, if we go back to the scientific, even commonsense point of view, implicitly time is taken as the ground of everything in scientific work. In fact even in ancient Greek mythology Chronos, the god of time, produces his children and swallows them. That is exactly what we said about the ground; everything comes from the ground and dies to the ground. So in a way, mankind long ago began to take time already as the ground.

JK: Yes. And then someone comes along and says time is not the ground.

DB: That's right. So until now even scientists have been looking for the ground somewhere in time—and everybody else too!

JK: That is the whole point.

DB: Now, you say time is not the ground. Somebody might say this is nonsense, but we say we will stay open to that, although some people might easily dismiss it right away. Now, if you say time is not the ground, we don't know where we are.

JK: I know where I am. We will go into it.

Q: Is time the same movement as this thought which we described first?

JK: Yes, time is that. Time is thought.

DB: Let's go slowly again on that, because there is, as we have often said, chronological time.

JK: Of course, that is simple.

DB: Yes, but in addition we are thinking. You see, thinking takes time chronologically, but in addition it projects a kind of imaginary time . . .

JK: . . . which is the future.

DB: Which is the future and the past as we experience it.

JK: Yes, that is right.

DB: That time which is imagined is also a kind of real process of thinking.

JK: It is a fact.

DB: It is a fact that it takes time, physically, to think, but we also have time when we can imagine the whole past and future.

JK: Yes, which are facts.

DB: So let's say that this time is not the ground, perhaps not even physically.

JK: We are going to find out.

DB: Yes, but we feel it to be the ground, because we feel that we, as the self, exist in time. Without time there could be no "me."

JK: That's it.

DB: "I" must exist in time.

JK: Of course, of course.

DB: Eternally being something or becoming something.

JK: Becoming and being are in the field of time. Now, can the mind, which has evolved through time . . .

Q: What do you mean by "mind" then?

JK: Mind—the brain, my senses, my feeling, all that is the mind.

DB: The particular mind, you mean.

JK: Particular mind, of course, I am talking of the mind that has evolved through time.

DB: Even its particularity depends on time.

JK: Time, of course, and all the rest of it. Now we are asking, can that mind be free of time to have an insight which is totally rational, which then can operate on thought? *That* thought is totally rational, not based on memory. Agreed?

DB: Yes.

JK: Now, how am I—X—to be free of time? I know I need time to go from here to there, to learn a lesson, a technique, etc. I understand that very clearly, so I am not talking about that time. I am talking about time as becoming.

DB: As being.

JK: Of course, becoming is being. I start from being to become.

DB: And being something in myself. Being better, being happier.

JK: Yes, the whole thing—the more. Now, can I, can my brain investigating to find out if the ground exists, can my whole mind be free of time? We have now separated time—the time which is necessary, and the time which is not necessary. That is, can my brain not function, as it has always done, in time as thought? Which means can thought come to an end? Would you accept that?

DB: Yes, but could you make that more clear? We can see that the first question is can my brain not be dominated by the function of thought?

JK: Yes, which is time.

DB: And then, if you say thought comes to an end . . .

JK: No! Can time as thought come to a stop?

DB: The psychological time come to a stop?

JK: Yes, I am talking of that.

DB: But we will still have the rational thought.

JK: Of course. That is understood. We have said that.

DB: We are discussing the thought of conscious experience.

Q: Of becoming and being . . .

JK: And the retention of memory; you know, the past, as knowledge. Oh, yes, that can be done.

DB: You really mean the memory of experiences?

JK: The memory of experiences, hurts, attachments, the whole of it. Now, can that come to an end? Of course it can. This is the point: It can come to an end when the very perception asks, what is it? What is hurt? What is psychological damage? The perception of it is the ending of it. Not carrying it over, which is time. The very ending of it is the ending of time. I think that is clear. X is hurt, wounded from childhood. And he, by listening, talking, discussing, realizes that the continuation of the hurt is time. And to find out the ground, time must end. So he says, can my hurt end instantly, immediately?

DB: Yes, I think there are some steps in that. You say, he finds that hurt is time, but the immediate experience of it is that it exists on its own.

JK: I know, of course. We can go into that.

DB: That it simply is something on its own.

JK: Which means I have created an image about myself and the image is hurt, but not me.

DB: What do you mean by that?

JK: All right. In the becoming, which is time, I have created an image about myself.

DB: Well, thought has created that image.

JK: Thought has created an image through experience, through education, through conditioning, and made this image separate from me. But this image is actually "me," although we have separated the image and the me, which is irrational. So in realizing that the image is "me," I have become somewhat rational.

DB: I think that will not be clear—because if I am hurt, I feel the image is "me."

JK: The image *is* you.

DB: The person who is hurt feels that way, you see.

JK: All right. But the moment you operate on it, you separate yourself.

DB: That's the point. Now, the first feeling is that the image is "me" hurt, and the second feeling is that I draw back from the image in order to operate on it . . .

JK: . . . which is irrationality.

DB: . . . because it is not correct.

JK: That's right.

DB: And that brings in time, because I say it will take time to do that.

JK: Quite right. So by seeing that, I become rational and act. The act is to be free of it immediately.

DB: Let's go into that. The first thing is that there has been a hurt. That is the image, but at first I don't separate it. I feel identified with it.

JK: I am that.

DB: I am that. But then I draw back and say that I think there must be a "me" who can do something.

JK: Yes, can operate on it.

DB: Now, that takes time.

JK: That is time.

DB: That is time, but I mean I am thinking it takes time. Now, I have to go slowly. If I don't do that, that hurt cannot exist.

JK: That's right.

DB: But it is not obvious in the experience itself that this is so.

JK: First let's go slowly into it. I am hurt. That is a fact. Then I separate myself—there is a separation—saying I will do something about it.

DB: The "me" who will do something is different.

JK: Different, of course.

DB: And he thinks about what he should do.

JK: The "me" is different because it is becoming.

DB: It projects into the future a different state.

JK: Yes. I am hurt. There is a separation, a division. The "me," which is always pursuing the becoming, says, I must control it. I must wipe it out. I must act upon it, or I will be vengeful, hurtful. So this movement of separation is time.

DB: We can see that now. The point is there is something here that is not obvious. A person is thinking that the hurt exists independently of "me," and I must do something about it. I project into the future the better state and what I will do. Let's try to make this very clear, because you are saying that there is no separation.

JK: My rationality discovers there is no separation.

DB: There is no separation, but the illusion that there is a separation helps to maintain the hurt.

JK: That's right. Because the illusion is "I am becoming."

DB: Yes. I am this and will become that. So I am hurt and I will become non-hurt. Now, that very thought maintains the hurt.

JK: That's right.

Q: Is the separation not already there when I become conscious and say I am hurt?

JK: I am hurt. Then I say, I am going to hit you because you have hurt me. Or I say, I must suppress it. Or I create fear, and so on.

Q: But isn't that feeling of separation there from the moment I say I am hurt?

JK: That is irrationality.

Q: That is irrational already?

JK: Yes, when you say, does not the separation exist already when I say, "I am hurt."

DB: It does, but I think that before that happens you get a kind of shock. The first thing that happens is a kind of shock, a pain or whatever, with which you are identified. Then you explain it by saying "I am hurt," and that immediately implies the separation to do something about it.

JK: Of course. If I am not hurt, I don't know anything about separation or not separation. If I am hurt, I am irrational as long as I maintain that hurt and do something about it, which is to become. Then irrationality comes in. I think that is right.

DB: Now, if you don't maintain it, what happens? Suppose you say, I won't go on with this becoming?

JK: Ah, that is quite a different matter. It means I am no longer observing with time, or using time as an observation.

DB: You could say that is not your way of looking. It is not your theory anymore.

JK: That's right.

DB: Because you could say time is a theory which everybody adopts for psychological purposes.

JK: Yes. That is the common factor; time is the common factor of man. And we are pointing out time is an illusion . . .

DB: Psychological time.

JK: Of course, that is understood.

DB: Are you saying that when we no longer approach this through time, then the hurt does not continue?

JK: It does not continue; it ends—because you are not becoming anything.

DB: In becoming you are always continuing what you are.

JK: That's right. Continuing what you are, modified . . .

DB: That is why you struggle to become.

JK: We are talking about insight. That is, insight has no time. Insight is not the product of time, time being memory, etc. So there is insight. That insight being free of time acts upon memory, acts upon thought. That is, insight makes thought rational, but not thought which is based on memory. Then what the devil is that thought?

No, wait a minute. I don't think thought comes in at all. We said insight comes into being when there is no time. Thought—which is based on memory, experience, knowledge—that is the movement of time as becoming. We are talking of psychological and not chronological time. We are saying to be free of time implies insight. Insight, being free of time, has no thought.

DB: We said that it may use thought.

JK: Wait. I am not sure. Just go slowly. It may use thought to explain, but it acts. Before, action was based on thought. Now, when there is insight, there is only action. Why do you want thought? Because insight is rational, action is rational. Action becomes irrational when it is acting from thought. So insight doesn't use thought.

DB: Well, we have to make it clear because in a certain area it has to use thought. . . . If, for example, you want to construct something, you would use the thought which is available on how to do it.

JK: But that is not insight.

DB: But even so you may have to have insight in that area.

JK: Partial. The scientists, the painters, the architects, the doctors, the artists, and so on have partial insight. But we are talking of X and Y and Z, who are seeking the ground; they are becoming rational, and we are saying insight is without time, and therefore without thought, and that insight is action. Because that insight is rational, action is rational. Forgive me, I am not making myself an example. I am talking in all humility. That boy, that young man in 1929 dissolved the Order of the Star. There was no thought. People said, "Do this," "Don't do that," "Keep it," "Don't keep it." He had an insight, dissolved it. Finished! Why do we need thought?

DB: But then you used some thought in dissolving the Order to say when to do it, how to do it.

JK: That is merely for convenience, for other people.

DB: But still some thought was needed.

JK: The decision acts.

DB: I didn't mean about the decision. The primary action did not require thought; those which follow may.

JK: That is nothing. It is like moving a cushion from there to there.

DB: Yes, I understand that. Then the primary source of action does not involve thought.

JK: That is all I wanted to say.

DB: But it sort of filters through into . . .

JK: It is like a wave.

Q: Does not all thought undergo a transformation in this process?

JK: Yes, of course. Because insight is without time; therefore the brain itself has undergone a change.

DB: Yes. Now could we talk about what you mean by that?

JK: Does it mean that every human response must be viewed by or must enter into insight? I will tell you what I mean. I am jealous. Is there an insight which will cover the whole field of jealousy and so end it? End envy, greed, and all that is involved in jealousy. You follow? Irrational people go step by step—get rid of jealousy, get rid of attachment, get rid of anger, get rid of this, that, and the other. Which is a constant process of becoming, right? But insight, which is totally rational, wipes all that away.

DB: Right.

JK: Is that a fact? Fact, in the sense that X, Y, and Z will never be jealous again. Never!

DB: We have to discuss that, because it is not clear how you could guarantee that.

JK: Oh, yes, I will guarantee it!

DB: If it can reach those who are able to listen ...

JK: Which means that to find the ground the first thing is to listen.

DB: You see, scientists cannot always listen. Even Einstein and Bohr were not able at a certain point to listen to each other. Each one was attached to his particular view.

JK: They brought their irrationality into operation.

Four

Breaking the Pattern of
Egocentric Activity

10 APRIL 1980, OJAI, CALIFORNIA

JIDDU KRISHNAMURTI: I would like to ask a question which may lead us to something: What will make man, a human being, change, deeply, fundamentally, radically? He has had crisis after crisis, he has had a great many shocks, he has been through every kind of misfortune, every kind of war, personal sorrow, and so on. A little affection, a little joy, but all this doesn't seem to change him. What will make a human being leave the way he is going and move in a totally different direction? I think that is one of our great problems, don't you? Why? If one is concerned, as one must be, with humanity, with all the things that are going on, what would be the right action to move man out of one direction to another? Is this question valid? Has it any significance?

DAVID BOHM: Well, unless we can see this action, it won't have much significance.

JK: Has the question any significance?

DB: What it means is, indirectly, to ask what is holding people.

JK: Yes—same thing.

DB: If we could find out what is holding people in their present direction ...

JK: Is it the basic conditioning of man, this tremendous egotistic attitude and action, which apparently won't yield to anything? It appears to change, it appears to yield, but the centre remains the same. Perhaps this may not be in the line of our dialogue over the last two or three days, but I thought we might start with this.

DB: Have you some notion of what is holding people? Something that would really change them?

JK: I think so.

DB: What is it then?

JK: What is it that is blocking? Do we approach it through environmental conditioning, from the outer to the inner, and discover, from man's outer activities, the inner? And then discover that the outer is the inner, the same movement, and then go beyond it to see what it is? Could we do that?

DB: When you say outward, what do you mean? Do you mean the social conditions?

JK: The social conditioning, the religious conditioning, education, poverty, riches, climate, food—the outer. Which may condition the mind in a certain direction. But as one examines it a little more, the psychological conditioning is also somewhat from the outer.

DB: It is true that the way a person thinks is going to be affected by his whole set of relationships. But that doesn't explain why the conditioning is so rigid and why it holds.

JK: That is what I am asking too.

DB: Yes. If it were merely outward conditioning, one would expect it to be more easily changed. For example, you could have some other outward condition.

JK: They have tried all that.

DB: Yes, the whole belief of communism was that with a new society there would be a new man.

JK: But there have been none!

DB: I think that there is something fundamentally in the nature of the inward that holds, that resists change.

JK: What is it? Will this question lead us anywhere?

DB: Unless we actually uncover it, it will lead nowhere.

JK: I think one could find out, if one applied one's mind. I am just asking, is this question worthwhile, and is it related to what we have been discussing? Or shall we take up something else in relation to what we have been talking about?

DB: Well, I think that we have been talking of bringing about an ending to time, an ending to becoming. And we talked of coming into contact with the ground, through complete rationality. But now we could say that the mind is not rational.

JK: Yes, we said man is basically irrational.

DB: This is perhaps part of the block. If we were completely rational, then we would of necessity come to this ground. Would that be right?

JK: Yes. We were talking the other day about the ending of time. The scientists, through the investigation of matter, want to find out that point. Also the so-called religious people have endeavoured to find out—not only verbally—if time can stop. We went into that quite a bit, and we say it *is* possible for a human being who will listen to find out through insight the ending of time. Because insight is not memory. Memory is time, memory is knowledge stored up in the brain, and so on. As long as that is in operation, there is no possibility of having insight into anything. Total insight, not partial insight. The artist, the musician, they all have partial insights and therefore they are still time-bound.

Is it possible to have a total insight, which is the ending of the "me," because the "me" is time? Me, my ego, my resistance, my hurts, all that. Can that "me" end? It is only when that ends that there is total insight. That is what we discovered.

And we went into the question, is it possible for a human being to end totally this whole structure of the "me"? We said yes and went into it. Very few people will listen to this because it is perhaps too frightening. And the question then arises: If the "me" ends, what is there? Just emptiness? There is no interest in that. But if one is investigating without any sense of reward or punishment, then there is something. We say that something is total emptiness, which is energy and silence. Well that sounds nice, but it has no meaning to an ordinary man who is serious and wants to go beyond it, beyond himself. And we pushed it further: Is there something beyond all this? And we said there is.

Q: The ground.

JK: The ground. Is it that the beginning of this enquiry is to listen? Will I, as a human being, give up my egocentric activity completely? What will make me move away from that? What will make a human being move away from this destructive, self-centred activity? If he will move away through reward, or punishment, then that is just another thought, motive. So discard that. Then what will make human beings renounce—if I may use the word—renounce it, without reward, completely?

You see, man has tried everything in this direction—fasting, self-torture in various forms, abnegating himself through belief and denying himself through identification with something greater. All the religious people have tried this, but the "me" is still there.

DB: Yes. The whole activity has no meaning, but somehow this does not become evident. People will move away from something which has no meaning and makes no sense, ordinarily speaking. But it seems that the perception of this fact is rejected by the mind. The mind is resisting it.

JK: The mind is resisting this constant conflict and moving away from it.

DB: It is moving away from the fact that this conflict has no meaning.

JK: People don't see that.

DB: Also the mind is set up purposefully to avoid seeing it.

JK: The mind is avoiding it.

DB: It is avoiding it almost on purpose, but not quite consciously, like the people in India who say they are going to retire to the Himalayas because nothing can be done.

JK: But that is hopeless. You mean to say that the mind, having lived so long in conflict, refuses to move away from it?

DB: It is not clear why it refuses to give it up, why the mind does not wish to see the full meaninglessness of the conflict. The mind is deceiving itself; it is continually covering up.

JK: The philosophers and so-called religious people have emphasized struggle, emphasized the sense of striving, control, effort. Is that one of the causes why human beings refuse to let go of their way of life?

DB: Possibly. They hope that by fighting or struggling they will achieve a better result. Not to give up what they have, but to improve it by struggle.

JK: Man has lived for two million years; what has he achieved? More wars, more destruction.

DB: What I am trying to say is that there is a tendency to resist seeing this, to continually go back to hoping that the struggle will produce something better.

JK: I am not quite sure if we have cleared this point, that the intellectuals—I am using the word respectfully—the intellectuals of the world have emphasized this factor of struggle.

DB: Many of them have, I suppose.

JK: Most of them.

DB: Karl Marx.

JK: Marx and even Bronowski, who talk of more and more struggle, of acquiring more and more knowledge. Is it that the intellectuals have such extraordinary influence on our minds?

DB: I think people do this without any encouragement from intellectuals. You see, struggle has been emphasized everywhere.

JK: That is what I mean. Everywhere. Why?

DB: Well, in the beginning people thought it would be necessary because they had to struggle against nature in order to survive.

JK: So struggling against nature has been transferred to the other?

DB: Yes, that is part of it. You see, you must be a brave hunter, and you must struggle against your own weakness to become brave. Otherwise you can't do it.

JK: Yes, that's it. So is it that our minds are conditioned, shaped, held, in this pattern?

DB: Well, that is certainly true, but it doesn't explain why it is so extraordinarily hard to change it.

JK: Because I am used to it. I am in a prison, but I am used to it.

DB: But I think that there is a tremendous resistance to moving away from it.

JK: Why does a human being resist this? If you come along and point out the fallacy, the irrationality of this, and you show the whole cause and effect, give examples, data, everything else? Why?

DB: That is what I said, that if people were capable of complete rationality they would drop it, but I think that there is something more to the problem. You see, you may expose the irrationality of it, but there is something more, in the sense that people are not fully aware of this whole pattern of thought. Having had it exposed at a certain level, it still continues at levels that they are not aware of.

JK: But what would make them aware?

DB: That is what we have to find. I think people have to become aware that they have this tendency to go on with the conditioning. It might be mere habit, or it might be the result of many past conclusions, all operating *now*, without people knowing it. There are so many different things that keep people in this pattern. You might convince somebody that the pattern makes no sense, but when it comes to the actual affairs of life, he has a thousand different ways of proceeding which imply that pattern.

JK: Quite. Then what?

DB: Well, I think that a person would have to be extremely interested in this to break all that down.

JK: Then what will bring a human being to this state of extreme interest? You see, they have even been offered heaven as a reward if they do this. Various religions have done this, although that becomes too childish.

DB: That is part of the pattern: reward. Ordinarily the rule is that I follow the self-enclosed pattern except when something really big comes up.

JK: A crisis.

DB: Or when a great reward is to be obtained.

JK: Of course.

DB: That is a pattern of thinking. People must in some way believe that it has value. If everybody were able to work together and suddenly we were able to produce harmony, then everybody would say, fine, I will give up myself to it. But in the absence of that, I had better hold on to what I have! That is the sort of thinking.

JK: Hold on to what is known.

DB: I don't have much, but I had better hold on to it.

JK: Yes. So are you saying that if everybody does this, I will do it?

DB: That is the common way of thinking. Because as soon as people begin to start to cooperate in an emergency, then a great many people go along.

JK: So they form communes. But all those have failed.

DB: Because after a while this special thing goes away and they fall back to the old pattern.

JK: The old pattern. So I am asking, what will make a human being break through this pattern?

QUESTIONER: Isn't it related to the question we dealt with before—time and no time?

JK: But I know nothing about time, I know nothing about all this; it is just a theory to me. Yet the fact is I am caught in this pattern and I can't let it go. The analysts have tried it, the religious people have tried it, everybody intelligent has tried to make human beings let this go, but apparently they have not succeeded.

Q: But they don't see that the very attempt at letting go the pattern or ending the conflict is still strengthening the conflict.

JK: No, that is just a theory.

Q: But you can explain that to them.

JK: You can explain. As we said, there are a dozen very rational explanations. At the end of it, we fall back to this.

Q: Well, you only fall back to that if you have not really understood it.

JK: Have *you* understood it when you say that? Why haven't I, or you, said "finished"? You can give me a thousand explanations and all probably a bit rational, but I say, have you done it?

Q: I don't even understand the question, when you ask have I done it.

JK: I am not being personal. You have given an explanation of why human beings can't move away from this pattern or break through it.

Q: No, I give you more than the explanation.

JK: What do you give me?

Q: If I observe something to be correct, then the description of the observation is more than just explanation.

JK: Yes, but can I observe this clearly?

Q: Well, that is the problem.

JK: So help me to see it clearly.

Q: For that there must be an interest.

JK: Please don't say "must." I haven't got an interest. I am interested, as Dr. Bohm pointed out just now, when there is a tremendous crisis such as war. Then I forget myself. In fact, I am glad to forget myself, to give the responsibility to the generals, to the politicians. In a crisis I forget, but the moment the crisis goes away I am back to my pattern. That is happening all the time. Now, I say to myself, what will make me relinquish this pattern or break through it?

Q: Isn't it that one must see the falseness?

JK: *Show* it to me.

Q: I can't, because I have not seen it.

JK: Then what shall I do as a human being? You have explained to me ten thousand times how ugly it is, how destructive it is, and so on, but I fall back to this pattern all the time. Help me, or show me how to break the pattern. You understand my question?

Q: Well, then you are interested.

JK: All right. Now, what will make me be interested? Pain?

Q: Sometimes it does for a moment, but it goes away.

JK: So what will make me as a human being so alert, so aware, so intense that I will break through this thing?

Q: You state the question in terms of an action, breaking through, relinquishing. Isn't it a matter of seeing?

JK: Yes. Show me, help me to see, because I am resisting you. My pattern, so deeply ingrained in me, is holding me back, right? I want proof; I want to be convinced.

Q: We have to go back to this question: Why do I want to have proof? Why do I want to be convinced?

JK: Because someone says that this is a stupid, irrational way of living. And he shows us all the effects of it, the cause of it, and we say, yes, but we can't let go!

DB: You may say that is the very nature of "me," that I must fulfil my needs no matter how irrational they are.

JK: That is what I am saying.

DB: First I must take care of my own needs, and then I can try to be rational.

JK: What are our needs then?

DB: Some of the needs are real and some are imaginary, but . . .

JK: Yes, that's it. The imaginary, illusory needs sway the other needs.

DB: But, you see, I may need to believe I am good and right and to know that I will be always there.

JK: Help me to break that!

DB: I think I have to see that this is an illusion. You see, if it seems real, what can I do? Because if I am really there, I need all this, and it is foolish to talk of being rational if I am going to vanish, break down, or something. You have proposed to me that there is another state of being where I am not there, right? And when I am there, this doesn't make any sense!

JK: Yes, quite. But I am *not* there. Suppose as a human being, heaven is perfect, but I am not there; please help me to get there.

DB: No, it is something different.

JK: I know what you are saying.

Q: Can one see the illusory nature of that very demand that I want to go to heaven? Or I want to be enlightened, or I want to be this, I want to be that? But this very question, this very demand is . . .

JK: This demand is based on becoming, on the more.

Q: That is illusory.

JK: No. You say that.

DB: You haven't demonstrated it to me, you see.

JK: That is an idea to you. It is just a theory. Show me.

Q: Well, are we willing really to explore the question?

JK: We are willing on one condition: that we find something at the end of it. See how the human mind works. I will climb the highest mountain if I can get something out of it.

Q: Can the mind see that this is the problem?

JK: Yes, but it can't let go.

Q: Well, if it sees . . .

JK: You are going round and round in circles!

DB: It sees the problem abstractly.

JK: That is it. Now, why do I see it abstractly?

DB: First of all, it is a lot easier.

JK: Don't go back to that. Why does my mind make an abstraction of everything?

DB: Let's begin by saying that to a certain extent it is the function of thought to make abstractions outwardly, but then we carry them inwardly. It is the same sort of thing as before.

JK: Yes. So is there something else—I am just asking—that we are missing in this altogether? That is, if I may point out, we are still thinking in the same old pattern.

DB: The question itself contains that pattern, doesn't it?

JK: Yes, but the pursuit of the pattern is traditional.

DB: I mean that in framing this question, the pattern has continued.

JK: Yes, so can we move away altogether from this and look at it differently? Can the human mind say, all right, we have tried all this—Marx, Buddha, everybody has pointed out something or other. But obviously after a million years, we are still somehow caught in that pattern—saying we must be interested, we must listen, we must do this, and so on.

DB: That is still time.

JK: Yes. Then what happens if I leave all that, actually leave it? I won't even think in terms of it. No more explanations or new twists that are the same *old* twists! So I say let's leave that area completely and look at the problem differently, the problem being why do I always live in this centre of "me, me, me"? I am a serious human being; I have listened to all this, and after fifty years I know all the explanations—what I should, should not do, etc. Can I say, "All right, I will discard all that"? That means I stand completely alone. Does that lead anywhere?

DB: Possibly, yes.

JK: I think it *does* lead somewhere.

DB: It seems to me that basically you are saying, "Leave all this knowledge of mankind behind."

JK: That is what I am saying.

DB: Apparently it is out of its place.

JK: Yes. Leave all the knowledge and experiences, explanations, causes that man has created—discard all that.

Q: But you are still left with the same mind.

JK: Ah! I have not such a mind. It is not the same mind. When I discard all this, my mind has changed. My mind is *this*.

Q: No, isn't the mind also the basic set-up?

JK: Which I have discarded.

Q: But you can't discard that.

JK: Oh, yes.

Q: I mean, this is an organism.

JK: Now, wait a minute. My organism has been shaped by knowledge, by experience. And more knowledge which I have acquired as I have evolved, as I have grown. As I have gathered more and more, it has strengthened "me," and I have been walking on that path for millennia. And I say, perhaps I may have to look at this problem totally differently—which is not to walk on that path at all, but to discard *all* knowledge I have acquired.

DB: In this area, in this psychological place.

JK: Psychologically, of course.

DB: At the core, at the source, knowledge is irrelevant.

JK: Yes.

DB: Further down the line it becomes relevant.

JK: Of course. That is understood.

Q: But I have one question. The mind at the beginning of its evolution was in that same position. The mind at the beginning of whatever you call man was in that position. It didn't have any knowledge.

JK: No. I don't accept that. Why do you say that? The moment it comes into being, it is already caught in knowledge. Would you say that?

DB: I think it is implicit in the structure of thought.

JK: That is just it.

DB: First of all, to have knowledge about the outward, and then to apply this to the inward, without understanding that it was going to be caught in it. Therefore it extended that knowledge into the area of psychological becoming.

Q: Well, if the mind started anew, it would go through the same mistake again.

JK: No, certainly not.

Q: Unless it has learnt.

JK: No, I don't want to learn. You are still pursuing the same old path. I don't want to learn. Please, just let me go into this a little bit.

DB: We should clear this up because on other occasions you have said it is important to learn, even about observing yourself.

JK: Of course.

DB: Now you are saying something quite different. It should be made clear why it is different. Why is it that you have given up the notion of learning at this stage?

JK: At this stage, because I am still gathering memory.

DB: But there was a state when it was important to learn about the mind.

JK: Don't go back. I am just starting. I have lived for sixty, eighty, or a hundred years. And I have listened to all this—the teachers in India, the Christians, the Muslims; I have listened to all the psychological explanations, to Freud, to Marx, and everybody.

DB: I think we should go a bit further. We agree that is all negative stuff, but in addition perhaps I have observed myself and learned about myself.

JK: About myself—yes, add that.

Q: And add K.

JK: Add K. And, at the end of it, I say perhaps this is a wrong way of looking at it, right?

DB: Right. Having explored that way, we finally are able to see it might be wrong.

JK: Perhaps. Perhaps I'm just pushing . . .

DB: Well, I would say that in some sense perhaps it was necessary to explore that way.

JK: Or not necessary.

DB: It may not have been, but given the whole set of conditions, it was bound to happen.

JK: Of course. So now I have come to a point when I say discard—we will put in that word—all that knowledge, because that hasn't led me anywhere, in the sense that I am not free of my ego-centricism.

DB: But that alone isn't enough, because if you say it hasn't worked, you can always hope or suppose that it *may*. But in fact you could see that it *can't* work.

JK: It can't work. I am definite on that.

DB: It is not enough to say it hasn't worked; actually it *cannot* work.

JK: It cannot work because it is based on time and knowledge, which is thought. And these explanations are based on thought—to acquire knowledge and so on and so on. Would you say that?

DB: As far as we have gone we have based it on knowledge and thought. And not only thought but the habitual patterns of skill, which are an extension of thought.

JK: So I put those aside—not casually, not with an interest in the future, but seeing the same pattern being repeated and repeated; different colours, different phrases, different pictures, different images—I discard all that totally. Instead of going north, as I have been going for millennia, I have stopped and am going east, which means my mind has changed.

Q: Has the structure of the "me" gone?

JK: Obviously.

Q: Without insight into it?

JK: No. I won't bring in insight for the moment.

DB: But there was insight to do that. I mean to say that to consider doing it was an insight. The insight was the thing that worked.

IK: I won't bring in that word.

DB: When you said that the whole thing could not work, I think that is an insight.

JK: For me. I see it cannot work. But then we go back again to how do I acquire insight and all that.

DB: But leaving that aside and just saying that it was an insight, the question of how to acquire it is not the point.

JK: It is an insight that says "out."

Q: "Out" to the pattern?

JK: No, finished with this constant becoming through experience, knowledge, patterns. Finished!

Q: Would you say that the kind of thinking afterwards is a totally different kind of thinking? Evidently one still must think.

JK: I am not sure.

Q: Well, you may call it something else.

JK: Ah, I won't call it anything else. Please, I am just fishing around. After having lived a hundred years, I see everybody pointing out the way to end the self, and that way is based on thought, time, knowledge. And I say, sorry, I know all that. I have used that. I have an insight into that; therefore it falls away. Therefore the mind has broken the pattern completely. Not going north but east, you break the pattern.

Now, all right. Suppose Dr. Bohm has this insight and has broken away from the pattern. Please let us help another human being to come to that. Don't say you must be interested, you must listen, then fall back. You follow? What is your communication with another human being, so that he hasn't got to go through all this mess? What will make me absorb so completely what you have said, so that it is in my blood, in my brain, *everything*, so that I see this thing? What will you do? Or is there nothing to do? You follow? Because if you have that insight, it is a passion. It is not just a clever insight, nor is it possible to sit back and be comfortable. It is a passion that won't let you sit still; you must move, give—whatever it is. What will you do? You have that passion or this immense insight. And that passion must, like a river with a great volume of water flowing over the banks, move in the same way.

Now, I am a human being, ordinary, fairly intelligent, well read, experienced. I have tried this, that, and the other thing, and I meet someone who is full of this, and I say, why won't I listen to him?

Q: I think we do listen.

JK: Do we?

Q: Yes, I think so.

JK: Just go very, very slowly. Do we so completely listen that there is no resistance, no saying why, what is the cause, why should I? You follow what I mean? We have been through all that. We have walked the area endlessly, back and forward from corner to corner, north, south, east, west. And X comes along and says, look there is a different way of living, something totally new, which means listening *completely*.

Q: If there is a resistance, one does not see the resistance.

JK: Then go back to school. I am not being rude. Go back to school.

Q: What do you mean?

JK: Begin all over again on why you resist.

Q: But one doesn't see the resistance.

JK: Then I will show you your resistance, by talking. But yet you go back.

Q: Krishnaji,[3] did not your initial question go beyond this, when you asked, let's leave the listening, the rationality, the thought?

JK: Yes, but that is just an idea. Will you *do* it? X comes along and says, "Look, eat this."

Q: I would eat it if I could see it.

JK: Oh, yes, you can see it very clearly. We said, don't go back to the pattern. See! Then you say, how am I to see? Which is the old pattern. Just see! X refuses to enter that pattern.

Q: The pattern of explanation?

JK: Yes, knowledge, all that. He says come over, don't go back.

Q: Krishnaji, to talk about a normal situation in the world, there are a number of people who ask one with similar words to see, put thought aside; if one would really look at this thing, one would see it. That is what the priests tell us. So what is the difference?

---

3. In India, a suffix is placed after a person's name or title as a mark of respect.

JK: No, I am not a priest. I have left all that. I have left the church, the gods, Jesus, the Buddhas, the Krishnas; I have left all that, Marx, Engels, Lenin, Stalin [*laughs*], all the analysts, all the pundits, everybody. You see, you haven't done that. X says do that. Ah, you say, no, I can't do it until you show me there is something else beyond all that. And X says, "Sorry." Has that any meaning?

DB: Yes. I think that we say leave all the knowledge behind. But knowledge takes many subtle forms which we don't see.

JK: Of course. You are full of this insight and you have discarded all knowledge because of that. And another keeps on paddling over the pool of knowledge. And you say leave it. The moment we enter into explanations we are back in the game. And you refuse to explain.

You see, explanations have been the boat in which to cross to the other shore. And the man on the other shore says there is no boat. X says, "Cross!" He is asking something impossible, isn't he?

DB: If it doesn't happen right away, then it *is* impossible.

JK: Absolutely. He is asking something impossible for me to do. I am meeting X, who is immovable. Either I have to go round him, avoid him, or go over him. I can't do any of that. But X absolutely refuses to enter into the game of words. Then what am I, who have played games with words, to do? X won't leave me alone, in the sense that he may leave personally but I have met something immovable. And it is there night and day with me. I can't battle with it because there is nothing to get hold of.

So what happens to me when I meet something that is completely solid, immovable, absolutely true—what happens to me? Is that the problem? That we have never met something like that? We may climb the Himalayas, but Everest is always there. In the same way, perhaps human beings have never met something irrevocable. Something absolutely immovable. Either I am terribly puzzled by it or say, well, I can't do anything about it. Walk away from it. Or it is something that I must investigate—you follow—I must capture. Which is it [*laughs*]?

Here is a solid thing. I am confronted by it. As I said, I might run away from it, which I generally do. Or worship it. Or try to understand what it is. When I do all these things, I am back in the old pattern. So I discard that. When meeting X, who is immovable, I see what the nature of it is. I am movable, as a human being, but X is immovable. The contact with it does something; it must. It is not some mystic, occult stuff but it is simple, isn't it?

Q: Sir, it functions like a magnet, but it doesn't break something.

JK: No, because you haven't let go the pattern. It is not X's fault.

Q: I didn't say it was.

JK: No, the implication is that. Therefore you are back, you are dependent.

Q: What is taking place?

JK: I am saying, you meet X; what happens?

Q: You said an effort to understand.

JK: Ah, there you are, lost. You are back in the old pattern. You see it, you feel it, you know it, you recognize it. It doesn't matter what word you use; it is there.

DB: Well, can't you say that X communicates the absolute necessity of not going on with the old pattern, because you see it absolutely can't work.

JK: Yes, put it in your own words. All right.

DB: And therefore that is unalterable, immovable—is that what you mean?

JK: Yes, I am movable; X is immovable.

DB: Well, what is *behind* X, what is working in X is immovable. Wouldn't you say that?

JK: What is working is something of a shock at first, naturally. I have been moving, moving, moving, then I meet something that is immovable. Suddenly something takes place, obviously. You can see what takes place. X is not becoming, and I am becoming. And X has been through explanations and all the rest of it, and he shows that becoming is painful. I am putting it quickly, in a few words. And I meet that. So there is the sensitivity—all right, let's put it the other way. The explanations, the *discarding* of all the explanations has made me sensitive. Much more alert. When I meet something like X, naturally there is a response not in terms of explanation or understanding. There is a response to that. There is bound to be. Explanations have been given over and over again. I have listened, but either they have made me dull or I begin to see that explanations have no value at all. So in this process I have become extraordinarily sensitive to any word of explanation. I am allergic!

There is a danger in this too, because, you know, people have said when you go to the guru he gives; so be silent and you will receive. That's an illusion, you know. Well, I have said enough.

DB: I could just say that when one sees that this whole process of time and knowledge and so on won't work, then it stops. Now, this leaves one more sensitive, right?

JK: Yes, the mind has become sharp.

DB: All this movement was getting in the way.

JK: Yes, psychological knowledge has made us dull.

DB: It has kept the brain moving in an unnecessary way.

Q: All knowledge?

DB: Well, no. You could say in some sense that knowledge needn't make you dull, I suppose, if it starts from the clarity of where we don't have this knowledge at the core . . .

JK: Yes. You remember we said too, in our discussions, that the ground is not knowledge.

DB: You see, the first thing is it creates emptiness.

JK: That's it.

DB: But not yet the ground, not immediately the ground.

JK: That's right. You see, we have discussed all this; I hear it on the tape, it is printed in a book, and I say, yes, I get it. By reading it, I have explained, I have acquired knowledge. Then I say, I must have that.

DB: The danger is that there is great difficulty in communicating this in a book because that is too fixed.

JK: But that is what generally happens.

DB: But I think that the main point, which could communicate it, is to see that knowledge in all its forms, subtle and obvious, cannot solve the psychological problem; it can only make it worse. But then there is another energy which is involved.

JK: You see what is happening now? If any trouble arises, I go to a psychologist. In any family trouble, I go to somebody who will tell me what to do. Everything around me is being organized and making me more and more helpless. That is what is happening.

Five

The Ground of Being
and the Mind of Man

12 APRIL 1980, OJAI, CALIFORNIA

DAVID BOHM: Perhaps we could go further into the nature of the ground, whether we could come to it and whether it has any interest in human beings. And also whether there could be a change in the physical behaviour of the brain.

JIDDU KRISHNAMURTI: Could we approach this question from the point of view, why do we have ideas? And is the ground an idea? That is where we must first be clear. Why have ideas become so important?

DB: Perhaps because the distinction between ideas, and what is beyond ideas, is not clear. Ideas are often taken to be something more than ideas; we feel they are not ideas but a reality.

JK: That is what I want to find out. Is the ground an idea or is it imagination, an illusion, a philosophic concept? Or something that is absolute in the sense that there is nothing beyond it?

DB: How can you tell that there is nothing beyond it?

JK: I am coming to that, slowly. I want to see whether we look at that or perceive that or have an insight into that from a concept. Because after all the whole Western world—perhaps also the Eastern world—is based on concepts. The whole outlook, the religious beliefs, everything is based on that. But do we approach it from that point of view or as a philosophic investigation—philosophic, in the sense of love of wisdom, love of truth, love of investigation, the process of the mind? Are we doing that when we discuss, when we want to investigate, explain, or find out what that ground is?

DB: Well, perhaps not all the philosophers have been basing their approach on concepts, although certainly philosophy is taught through concepts. Certainly it is very hard to teach it except through concepts.

JK: What then is the difference between a religious mind and a philosophic mind? You understand what I am trying to convey? Can we investigate the ground from a mind that is disciplined in knowledge?

DB: Well, fundamentally, we say that the ground is unknown inherently. Therefore we can't begin with knowledge, and we have suggested we start with the unknown.

JK: Yes. Say for instance X says there is such a ground. And all of us, Y and Z, say what is that ground, prove it, show it, let it manifest itself. When we ask such questions, is it with a mind that is seeking or rather that has this passion, this love for truth? Or are we merely saying let's talk about it?

DB: I think that in that mind there is the demand for certainty: Show it; I want to be sure. So there is no enquiring.

JK: Suppose you state that there is such a thing, that there is the ground; it is immovable and so on. And I say I want to find out. Show it; prove it to me. How can my mind, which has evolved through knowledge, which has been highly disciplined in knowledge, even touch that? Because that is not knowledge; it is not put together by thought.

DB: Yes, as soon as we say prove it, we want to turn it into knowledge.

JK: That's it!

DB: We want it to be absolutely certain knowledge, so that there can be no doubt. And yet, on the other side of the coin, there is also the danger of self-deception and delusion.

JK: Of course. The ground cannot be touched as long as there is any form of illusion, which is the projection of desire, pleasure, or fear. So how do I perceive that thing? Is the ground an idea to be investigated? Or is it something that cannot be investigated?

DB: Right.

JK: Because my mind is trained, disciplined, by experience and knowledge, and it can only function in that area. And someone comes along and tells me that this ground is not an idea, is not a philosophic concept; it is not something that can be put together or perceived by thought.

DB: It cannot be experienced; it cannot be perceived or understood through thought.

JK: So what have I? What am I to do? I have only this mind that has been conditioned by knowledge. How am I to move away from all that? How am I, an ordinary man, educated, well-read, experienced, to feel this thing, to touch it, to comprehend it? You tell me words will not convey that. You tell me you must have a mind that is free from all knowledge, except that which is technological. And you are asking an impossible thing of me, aren't you? And if I say I will make an effort, then that also is born out of the self-centred desire. So what shall I do? I think that is a very serious question. That is what every serious person asks.

DB: At least implicitly. They may not say it.

JK: Yes, implicitly. So you, on the other side of the bank, as it were, tell me that there is no boat to cross in. You can't swim across. In fact you can't do anything. Basically, that is what it comes to. So what shall I do? You are asking me, you are asking the mind—not the general mind but . . .

DB: . . . the particular mind.

JK: You are asking this particular mind to eschew all knowledge. Has this ever been said in the Christian or the Jewish worlds?

DB: I don't know about the Jewish world, but in some sense the Christians tell you to give your faith to God, to give over to Jesus, as the mediator between us and God.

JK: Yes. Now, Vedanta means the end of knowledge. And being a Westerner, I say it means nothing to me. Because from the Greeks and all that, the culture in which I have lived has emphasized knowledge. But when you talk to some Eastern minds, they acknowledge in their religious life that a time must come when knowledge must end; the mind must be free of knowledge. But it is only a conceptual, a theoretical understanding. And to a Westerner, it means absolutely nothing.

DB: I think that there has been a Western tradition which is similar, but not as common. For example, in the Middle Ages there was a book called *The Cloud of Unknowing* which is on that line, although it is not the main line of Western thought.

JK: Not the main line. So what shall I do? How shall I approach the question? I want to find it. It gives meaning to life. It is not that my intellect gives meaning to life by inventing some illusion, some hope, some belief, but I see vaguely that this understanding, coming upon this ground, gives an immense significance to life.

DB: Well, people have used that notion of God to give significance to life.

JK: No, no. God is merely an idea.

DB: Yes, but the idea contains something similar to the Eastern idea that God is beyond knowing. Most people accept it that way, though some may not. So there is some sort of similar notion.

JK: But you tell me that the ground is not created by thought. So you cannot under any circumstances come upon it through any form of manipulation of thought.

DB: Yes, I understand. But I am trying to say that there is this problem, danger, delusion, in the sense that people say, "Yes, that is quite true; it is through a direct experience of Jesus that we come upon it, not through thought, you see!" I am not able to express their view accurately. Possibly the grace of God?

JK: The grace of God, yes.

DB: Something beyond thought, you see.

JK: As a fairly educated, thoughtful man, I reject all that.

DB: Why do you reject it?

JK: Because it has become common, first of all, common in the sense that everybody says that! And also there may be in it a great sense of illusion created by desire, hope, fear.

DB: Yes, but some people do seem to find this meaningful, although it may be an illusion.

JK: But if they had never heard of Jesus, they wouldn't experience Jesus.

DB: That seems reasonable.

JK: They would experience something different that they have been taught. In India, I mean . . .

QUESTIONER: But don't the more serious people in the religions say that essentially God, or whatever that is, the absolute, the ground, is something that cannot be experienced through thinking? Also they might go so far as to say it cannot be experienced at all.

JK: Oh, yes, I have said it cannot be experienced. X says it cannot be experienced. Now, let's say I don't know. Here is a person who says there is such a thing. And I listen to him, and not only does he convey it by his presence, but through the word. Although he tells me to be careful; the word is not the thing, but he uses the word to convey that there is this something so immense that my thought cannot capture it. And I say, all right, you have explained that very carefully, and how is my brain, that is conditioned, disciplined in knowledge, how is it to free itself from all that?

Q: Could it free itself by understanding its own limitation?

JK: So you are *telling* me thought is limited. Show it to me! Not by talking of memory, experience, or knowledge; I understand that, but I don't capture the feeling that it is limited, because I see the beauty of the earth, I see the beauty of a building, of a person, of nature. I see all that, but when you say thought is limited, I don't *feel* it. It is just a lot of words which you have said to me. Intellectually I understand. But I have no feeling for it. There is no perfume in it. How will you show me—not show me; how will you help me—not help—aid me, to have this feeling that thought itself is brittle, it is such a small affair? So that it is in my blood. You understand? When once it is in my blood, I have got it. You don't have to explain it.

Q: But isn't that the possible approach, not to talk about the ground—that at the moment is far too removed—but rather to look directly at what the mind can do.

JK: Which is thinking.

Q: The mind is thinking.

JK: That is all I have. Thinking, feeling, hating, loving—you know all that. The activity of the mind.

Q: Well, I would say we don't know it; we only *think* we know it.

JK: I know when I am angry. I know when I am wounded. It is not an idea; I have got the feeling. I am carrying the hurt inside me. I am fed up with the investigation because I have done it all my life. I go to Hinduism, Buddhism, Christianity, Islam, and I say I have investigated, studied, looked at them. I say these are all just words. How do I as a human being have this extraordinary feeling about it? If I have no passion, I am not investigating. I want to have this passion that will explode me out of this little enclosure. I have built a wall around myself, a wall which is myself. And man has lived with this thing for millions of years. And I have been trying to get out of it by studying, by reading, by going to gurus, by all kinds of things, but I am still anchored there. And you talk about the ground, because you see something that is breathtaking, that seems so alive, so extraordinary. And I am here, anchored in here. You, who have "seen" the ground, must do something that will explode, break up this centre completely.

Q: *I* must do something or *you* must?

JK: Help me! Not by prayer and all that nonsense. You understand what I am trying to say? I have fasted, I have meditated, I have renounced, I have taken a vow of this and that. I have done all those things. Because I have had a million years of life. And at the end of the million years, I am still where I was, at the beginning. This is a great discovery for me; I thought I had moved on from the beginning, by going through all this, but I suddenly discover I am back at the same point where I started. I have had more experience, I have seen the world, I have painted, I have played music, I have danced. You follow? But I have come back to the original starting point.

Q: Which is me and not me.

JK: *Me.* I say to myself, what am I to do? And what is the human mind's relationship to the ground? Perhaps if I could establish a relationship it might break up this centre, totally. This is not a motive, not a desire, not a reward. I see that if the mind could establish a relationship with *that,* my mind has become that, right?

Q: But hasn't mind then already become that?

JK: Oh, no.

Q: But I think you have just wiped away the greatest difficulty in saying there is no desire.

JK: No, no. I said I have lived a million years . . .

Q: But that is an insight.

JK: No. I won't accept insight so easily as that [*laughs*].

Q: Well, let me put it this way: It is something much more than knowledge.

JK: No, you are missing my point. My brain has lived for a million years. It has experienced everything. It has been Buddhist, Hindu, Christian, Muslim; it has been all kinds of things, but the core of it is the same. And someone comes along and says, look, there is a ground which is . . . something! Am I going back to what I have already known, the religions and so on? I reject all that, because I say I have been through it all, and they are like ashes to me at the end of it.

DB: Well, all those things were the attempt to create an apparent ground by thought. It seemed that through knowledge and thought, people created what they regarded as the ground. And it wasn't.

JK: It wasn't. Because man has spent a million years at it.

DB: So long as knowledge enters the ground, that will be false?

JK: Of course. So is there a relationship between that ground and the human mind? In asking that question, I am also aware of the danger of such a question.

DB: Well, you may create a delusion of the same kind that we have already gone through.

JK: Yes. I have played that song before.

Q: Are you suggesting that the relationship cannot be made by you, but it must come . . . ?

JK: I am asking that. No, it may be that I have to make a relationship. My mind now is in such a state that I won't accept a thing. My mind says I have been through all this before. I have suffered, I have searched, I have looked, I have investigated, I have lived with people who are awfully clever at this kind of thing.

So I am asking the question, being fully aware of the danger of it, as when the Hindus say God is in you, Brahman is in you, which is a lovely idea! But I have been through all that.

So I am asking if the human mind has no relationship to the ground, and if there is only a one-way passage, from that to me . . .

DB: Surely that's like the grace of God then, that you have invented.

JK: That I won't accept.

DB: You are not saying the relationship is one way, nor are you saying it is *not* one way.

JK: Maybe. I don't know.

DB: You are not saying anything.

JK: I am not saying anything. All that I "want" is this centre to be blasted. You understand? For the centre not to exist. Because I see that the centre is the cause of all the mischief, all the neurotic conclusions, all the illusions, all the endeavour, all the effort, all the misery—everything is from that core. After a million years, I haven't been able to get rid of it; it hasn't gone. So is there a relationship at all? What is the relationship between goodness and evil? Consider it. There is no relationship.

DB: It depends on what you mean by relationship.

JK: Contact, touch, communication, being in the same room . . .

DB: . . . coming from the same root.

JK: Yes.

Q: But are we then saying that there is the good and there is the evil?

JK: No, no. Let's use another word: "whole," and that which is not whole. It is not an *idea*. Now, is there relationship between these two? Obviously not.

DB: No, if you are saying that in some sense the centre is an illusion. An illusion cannot be related to that which is true, because the content of the illusion has no relation to what is true.

JK: That's it. You see, that is a great discovery. I want to establish relationship with that. "Want": I am using rapid words to convey something. This petty little thing wants to have relationship with that immensity. It cannot.

DB: Yes, not just because of its immensity, but because in fact this thing is not actual.

JK: Yes.

Q: But I don't see that. He says the centre is not actual, but I don't see that the centre is not actual.

DB: Not actual, in the sense of not being genuine but an illusion. I mean, something is acting but it is not the content which we know.

JK: Do you see that?

Q: You say the centre must explode. It does not explode because I don't see the falseness in it.

JK: No. You have missed my point. I have lived a million years; I have done all this. And at the end of it I am still back at the beginning.

Q: So you say the centre must explode.

JK: No, no, no. The mind says this is too damn small. And it can't do anything about it. . . . It has prayed; it has done everything. But the centre is still there. And someone tells me there is this ground. I want to establish a relationship with that.

Q: He tells me there is this thing and also says that the centre is an illusion.

DB: Wait. That is too quick.

JK: No. Wait. I know it is there. Call it what you like, an illusion, a reality, a fixation—whatever you like. It is there. And the mind says it is not good enough; it wants to capture that. It wants to have relationship with it. And that says, "Sorry, you can't have relationship with me." That's all!

Q: Is that mind which wants to be in connection, in relationship with that, the same mind which is the "me"?

JK: Don't split it up, please. You are missing something. I have lived all this. I know, I can argue with you, back and forth. I have a million years of experience, and it has given me a certain capacity. And I realize at the end of it all there is no relationship between me and truth. And that's a tremendous shock to me. It is as if you have knocked me out, because my million years of experience say go after that, seek it, pray for it, struggle for it, cry, sacrifice for it. I have done all that. And suddenly it is pointed out that I cannot have relationship with that. I have shed tears, left my family, *everything* for that. And *that* says, "No relationship." So what has happened to me? This is what I want to get at. Do you understand what I am saying, what has happened to me? To the mind that has lived this way, done everything in search of that, when that says, "You have no relationship with me." This is the greatest thing . . .

Q: It is a tremendous shock to the "me," if you say that.

JK: Is it to you?

Q: I think it was, and then . . .

JK: Don't! I am asking you, is it a shock to discover that your brain and your mind, your knowledge, are valueless? All your examinations, all your struggles, all the things that you have gathered through years and years, centuries, are absolutely worthless? Do you go mad, because you say you have done all this for nothing? Virtue, abstinence, control, everything—and at the end of it, you say they are valueless! Do you understand what this does to you? You don't see it.

DB: I mean, if the whole thing goes, then it is of no consequence.

JK: Absolutely, you have no relationship. What you have done or not done is absolutely of no value.

DB: Not in any fundamental sense. It has relative value, relative value only within a certain framework, which in itself has no value.

JK: Yes, thought has relative value.

DB: But the framework in general has no value.

JK: That's right. The ground says, "Whatever you have done on earth has no meaning." Is that an idea? Or an actuality? Idea being that you have told me, but I still go on, struggling, wanting, groping. Or is it an actuality, in the sense that I suddenly realize the futility of all that I have done. [*Long pause*] So one must be very careful to see that it is not a concept; or rather that one doesn't translate it into a concept or an idea but receives the full blow of it!

Q: You see, Krishnaji, for hundreds of years, probably since man has existed, he has pursued what he calls God, or the ground.

JK: As an idea.

Q: But then the scientific mind came along and also said it is just an idea, it is just foolish.

JK: Oh, no! The scientific mind says that through investigating matter we will perhaps come upon the ground.

DB: Yes, many feel that way. Some would even add investigate the brain, you see.

JK: Yes. That is the purpose of investigating the mind, not to blast each other off the earth with guns. We are talking of "good scientists," not governmental scientists, but those who say we are examining matter, the brain and all that, to find out if there is something beyond all this.

Q: And many people, many scientists, would say that they have found the ground; the ground is empty, it is emptiness; it is an energy which is indifferent to man.

JK: Now, is that an idea or an actuality to them, which affects their life, their blood, their mind, their relationship with the world?

Q: I think it is just an idea.

JK: Then I am sorry, I have been through that. I was a scientist ten thousand years ago! You follow? I have been through all that. If it is merely an idea, we can both play at that game. I can send the ball to you, it is in your court, and you can send it back to me. We can play that. But I have finished with that kind of game.

DB: Because, in general, what people discover about matter does not seem to affect them deeply, psychologically.

JK: No, of course not.

DB: You might think that if they saw the whole unity of the universe they would act differently, but they don't.

Q: You could say that it has affected some of their lives. You see, the whole communist doctrine is built on the idea, which they think is a fact, that whatever is, is just a material process, which is essentially empty. So then man has to organize his life and society according to those dialectical principles.

JK: No, no, dialectical principles are opinion opposing another opinion—man hoping, out of opinions, to find the truth.

DB: I think we should leave this aside. There are different ways of looking at the meaning of the word "dialectic"—it also means to see reality as a flowing movement, not to see things as fixed but to see them in movement and interconnection. But I think you could say that whatever way people managed to look at it, after they saw this unity, it didn't fundamentally change their lives. In Russia, the same structures of the mind, if not worse, hold as elsewhere. And wherever people have tried this, it has not actually, fundamentally, affected the way they feel and think, and the way they live.

Q: You see, what I wanted to say is that the dismissal of the pursuit of the ground has not had any shocking effect on people.

JK: No! I am not interested. It has given me a tremendous shock to discover the truth, that all the churches, prayers, books, have absolutely no meaning, except how we can build a better society and so on.

DB: If we could manage to bring this point to order, then it would have great meaning—to build a good society. But as long as this disorder is at the centre, we can't use that in the right way. I think it would be more accurate to say that there is a great potential meaning in all that. But it does not affect the centre, and there is no sign that it has ever done so.

Q: You see, what I don't understand is that there are many people who in their life have never pursued what you call the ground.

JK: They are not interested.

Q: Well, I am not so sure. How would you approach such a person?

JK: I am not interested in approaching any person. All the works I have done—everything I have done—the ground says are valueless. And if I can drop all that, my mind is the ground. Then from *there* I move. From *there* I create society. Sorry!

DB: I think you could say that as long as you are looking for the ground somewhere by means of knowledge, then you are getting in the way.

JK: So to come back to earth, why has man done this?

DB: Done what?

JK: Accumulated knowledge. Apart from the necessity of having knowledge in certain areas, why has this burden of knowledge continued for so long?

DB: Because in one sense man has been trying to produce a solid ground through knowledge. Knowledge has tried to create a ground. That is one of the things that has happened.

JK: Which means what?

DB: It means illusion again.

JK: Which means that the saints, the philosophers, have educated me—in knowledge and through knowledge—to find the ground.

DB: To create a ground by using knowledge.

Q: You see, in a way, there used to be all these periods when mankind was caught in superstition. And knowledge was able to do away with that.

JK: Oh, no.

Q: To some extent it was.

JK: Knowledge has only crippled me from seeing truth. I stick to that. It hasn't cleared me of my illusions. Knowledge may be illusion itself.

Q: That may be, but it has cleared up some illusions.

JK: I want to clear up all the illusions that I hold, not some. I have got rid of my illusion about nationalism; I have got rid of illusion about belief, about Christ, about this, about that. At the end of it, I realize my mind is illusion. You see, to me, who have lived for a thousand years, to find that all this is absolutely worthless is something enormous.

DB: When you say you have lived for a thousand years, or a million years, does that mean, in a sense, that all the experience of mankind is . . . ?

JK: . . . is me.

DB: Is me. Do you feel that?

JK: I do.

DB: And how do you feel it?

JK: How do we feel anything? Wait a minute. I will tell you. It is not sympathy or empathy, it is not a thing that I have desired; it is a *fact*, an absolute, irrevocable fact.

DB: Could we share that feeling, perhaps? You see, that seems to be one of the steps that is missing, because you have repeated that quite often as an important part of the whole thing.

JK: Which means that when you love somebody, there is no "me"—it is love. In the same way, when I say I am humanity, it is so; it is not an idea, it is not a conclusion, it is part of me.

DB: Let's say it is a feeling that I have gone through all that, all that you describe.

JK: Human beings have been through all that.

DB: If others have gone through it, then I also have gone through it.

JK: Of course. One is not aware of it.

DB: No, we separate.

JK: If we admit that our brains are not my particular brain, but the brain that has evolved through millennia . . .

DB: Let me say why this doesn't communicate so easily: Everybody feels that the content of his brain is in some way individual, that *he* hasn't gone through all that. Let's say that somebody, thousands of years ago, went through science or philosophy. Now, how does that affect me? That is what is not clear.

JK: Because I am caught in this self-centred, narrow little cell, which refuses to look beyond. But you as a scientist, as a religious man, come along and tell me that your brain is the brain of mankind.

DB: Yes, and all knowledge is the knowledge of mankind. So that in some way we have all knowledge.

JK: Of course.

DB: Though not in detail.

JK: So you tell me that, and I understand what you mean, not verbally, not intellectually: It is so. But I come to that only when I have given up ordinary things, like nationality and so on.

DB: Yes, we have given up the division, and we can see that the experience is of all mankind.

JK: It is so obvious. You go to the most primitive village in India, and the peasant will tell you all about his problems, his wife, children, poverty. It is exactly the same thing, only he is wearing different clothes, trousers, kimono, or whatever! For X, this is an indisputable fact: It is so.

He says, all right, at the end of all this, of all these years, I suddenly discover that it is empty. You see, we don't accept it, we are too clever. We are so soaked with disputations and arguments and knowledge. We don't see a simple fact. We refuse to see it. And X comes along and says, "See it, it is there." Then the immediate machinery of thought begins: Prove it. Then they say, "Be silent." So I practise silence! I have done that for a thousand years. It has led nowhere.

So there is only one thing, and that is to discover that *all* that I have done is useless—ashes! You see, that doesn't depress one. That is the beauty of it. I think it is like the phoenix.

DB: Rising from the ashes.

JK: Born of ashes.

DB: In a way, it is freedom, to be free of all that.

JK: Something totally new is born.

DB: Now, what you said before is that the mind is the ground, it is the unknown.

JK: The mind? Yes. But not this mind.

DB: In that case, it is not the same mind.

JK: If I have been through all that and come to a point when I have to end all that, it is a new mind.

DB: That's clear. The mind is its content, and the content is knowledge, and without that knowledge, it is a new mind.

Six

Can Insight Bring About a
Mutation of the Brain Cells?

15 APRIL 1980, OJAI, CALIFORNIA

DAVID BOHM: You have said that insight changes the brain cells, and I wonder if we could discuss that?

JIDDU KRISHNAMURTI: As it is constituted, the brain has functioned in one direction: memory, experience, knowledge. It has functioned in that area as much as possible, and most people are satisfied with it.

DB: Well, they don't know of anything else.

JK: And also they have placed knowledge in supreme importance. If one is concerned with fundamental change, where does one begin? Suppose X feels he is going along a certain direction set by mankind. He has been going there century after century, and he asks himself what is radical change, if it is in the environment or in human relationships, if it is a sense of love, which is not in the area of knowledge. Where is it to begin? You understand my question? Unless there is some mutation taking place inside here, inside my mind . . . brain, I may *think* I have changed, but it is a superficial change and not a change in depth.

DB: Yes. What is implied there is that the present state of affairs involves not only the mind but also the nervous system and the body. Everything is set in a certain way.

JK: Of course. That is what I meant, the whole movement is set in a certain way. And along that pattern I can modify, adjust, polish a little more, a little less, and so on. But if a man is concerned with radical change, where is he to begin? As we said the other day, we have relied on the environment or society and various disciplines to change us, but I feel these are all in the same direction.

DB: Insofar as they all emanate from the way the mind and body are set, they are not going to change anything. There is a total structure involved which is in the brain, in the body, in the whole of society.

JK: Yes, yes. So what am I to do? What is X to do? And in asking this question, what is there to change?

DB: What exactly do you mean by "What is there to change?" What is to be changed?

JK: Yes, both: What is there to be changed and what is there to change? Basically, what is there to change? X sees that he can change certain things along this way, but to go much further than that, what is one to do? I am sure man has asked this question. You must have asked it. But apparently the mutation hasn't taken place. So what is X to do? He realizes the need for a radical revolution, a psychological revolution. He perceives that the more he changes, it is the same thing continuing; the more he enquires into himself, the enquiry remains the same; and so on. So what is there to change, unless X finds a way to change the brain itself?

DB: But what will change the brain?

JK: That's it. The brain has been set in a pattern for millennia! I think it is no longer what should I change. It is imperative that I change.

DB: So it is agreed that there must be a change, but the question is still, how can the brain change?

JK: One must come to that point. If this question is put to you as a scientist, or as a human being who is involved in science, what would your answer be?

DB: I don't think science can deal with it, because it doesn't go far enough. It can't possibly probe that deeply into the structure of the brain. Many questions are posited about the relationship of brain and mind, which science has not been able to resolve. Some people would say that there is nothing beyond the brain . . .

JK: Purely materialistic; I understand all that.

DB: If it is not materialistic, then for the moment science has very little to say about it. Perhaps some people would try to, but science generally has been most successful, most systematic, in dealing with matter. Any attempt to do otherwise is not very clear.

JK: So you tell X, that insight changes the brain cells and so on. My immediate response to that is how? Everybody asks that. It is not a matter of faith. It is not a matter of changing one pattern to another pattern. So you leave me without any direction, right? You leave me without any instrument that can penetrate this.

DB: Except that you are *implying* there is something beyond the brain, in putting that question. We don't *know*. The very statement implies that insight is somehow beyond the brain, else it couldn't change the brain.

JK: Yes. So how am I to capture it? Maybe I can't capture it . . .

DB: . . . but how will it come about? You are saying that something that is non-material can affect matter. This is the implication.

JK: I am not sure.

DB: I think that clearing this up would make more clear what your question is. It is somewhat puzzling if you don't.

JK: All that you have said to me is insight changes, brings about mutation in the brain. Now, you explain what insight is, which is not a result of progressive knowledge, not progressive time, not a remembrance, which, you point out, may be the real activity of the brain.

DB: All right. Let's put it differently. The brain has many activities, which include memory and all these that you have mentioned. In addition there is a more inward activity, but it is still the activity of the brain.

JK: Then it would be the same.

DB: You see, in putting this, something seems to be not quite clear.

JK: Yes. We must be very clear that it is not the result of progressive knowledge; it is not come by through any exercise of will.

DB: Agreed. I think people can generally see that insight comes in a flash; it does not come through will. Those who have considered it at all can see that. Also, that chemistry, drugs, will probably not bring it about.

JK: I think most people who are concerned see that. But how am I, as X, to have this insight? I see your logic, I see your reason.

DB: In some ways it may disturb people. It is not clear what the logic is, what is going to make this change in the brain. Is it something more than the brain, or is it something deeper in the brain? This is one of the questions.

JK: Of course.

DB: It is not quite clear logically.

QUESTIONER: Are you saying that there is a function of the brain which acts without reference to its content?

JK: Yes, to the past, to the content.

DB: This is a good question. Is there a function in the brain which is independent of the content? Which is not conditioned by the content, but that might still be a physical function?

JK: I understand. Is this the question? Apart from the consciousness with its content, is there in the brain an activity which is not touched by consciousness?

DB: By the content, yes.

JK: Content is the consciousness.

DB: Yes, but sometimes you use the word in another sense. Sometimes you imply that there could be another kind of consciousness. So if we call it "content," it would be more clear.

JK: All right. A part of the brain which is not touched by the content.

DB: Yes, this suggests that it may be possible for the brain to change. Either the brain is entirely controlled by its content, or in some way it is not that conditioned, it has some . . .

JK: That is a dangerous concept!

DB: But it is what you are saying.

JK: No. I see the danger of it. I see the danger of admitting to oneself that there is a part of the brain . . .

DB: . . . an activity . . .

JK: All right, an activity of the brain which is not touched by the content.

DB: It is a possible activity. It may be that it has not been awakened.

JK: It has not been awakened. That's right.

Q: But what is the danger?

JK: That is simple enough. The danger is that I am admitting there is God in me, that there is something superhuman, something beyond the content which therefore will operate on it, or that will operate in spite of it.

Q: But which part of the brain sees the danger?

JK: Let us go slowly. Which part of the brain sees the danger? Of course it is the content that sees the danger.

Q: Does it?

JK: Oh, yes, because the content is aware of all the tricks it has played.

DB: This is similar to many of the old tricks.

JK: Yes.

DB: Those tricks we have discussed before—the assumption of God within, the imagination of God within. There is a danger here obviously.

Q: But could the brain, seeing the danger, make that statement nevertheless? Because that statement might be pointing to the right direction.

DB: Even though it is dangerous, it may be necessary to do so; it may be on the right track.

JK: The unconscious, which is part of the content, may capture this, and say, "Yes"—so it sees the danger instantly.

Q: It sees its own trap.

JK: Yes, it sees the trap which it has created. So it avoids that trap. That is sanity: To avoid a trap is sanity. Is there an activity which is totally independent of the content? Then is that activity part of the brain?

DB: Is it a natural activity of the brain? Material in the brain.

JK: Which means what?

DB: Well, if there is such a natural activity, it could awaken somehow, and that activity could change the brain.

JK: But would you say it is still material?

DB: Yes. There could be different levels of matter, you see.

JK: That is what I am trying to get at. Right.

DB: But, you see, if you think that way, there could be a deeper level of matter which is not conditioned by the content. For example, we know matter in the universe is not conditioned by the content of our brains generally. There could be a deeper level of matter not conditioned in that way.

JK: So it would still be matter, refined or "super," or whatever; it would still be the content.

DB: Why do you say that? You see, you have to go slowly. Do you say that matter is content?

JK: Yes.

DB: Inherently? But this has to be made clear, because it is not obvious.

JK: Let's clear this up slowly. Let's discuss it. It's rather good, this. Thought is matter.

DB: Well, all right, thought is part of the content, part of the material process. Whether it exists independently as matter is not so clear. You can say water is matter; you can pour water from one glass to another, it has an independent substance. But it is not clear whether thought could stand as matter by itself, except with some other material substance like the brain in which it takes place. Is that clear?

JK: I don't quite follow.

DB: If you say water is matter, then it is clear. Now, if you say thought is matter, then thought must have a similar independent substance. You say air is matter, right? Or water is matter. Now, waves are not matter, they are just a process in matter. Is that clear?

JK: Yes. A wave is a process in matter.

DB: A material process. Is thought matter or is it a process in matter?

Q: May one ask, is electricity considered to be matter?

DB: Insofar as there are electron particles, it is matter, but it is also a movement of that, which is a process.

Q: So it is two things.

DB: Well, you can form waves of electricity and so on.

Q: Waves would be the matter, but not the electrical action.

DB: The electrical action is like the waves, but the electricity consists of particles.

JK: What is the question we are now asking?

DB: Is thought a material substance, or is it a process in some other material substance, like the brain?

JK: It is a material process in the brain.

DB: Yes, scientists would generally agree with that.

JK: Let's stick to that.

DB: If you say it is matter, they would become very puzzled.

JK: I see.

Q: It doesn't exist apart from the brain cells. It resides in the brain.

JK: That is, thought is a material process in the brain. That would be right. Then can that material process ever be independent?

DB: Independent of what?

JK: Independent of something that is not a material process. No, wait a minute. We must go slowly. Thought is a material process in the brain. We all agree about this?

DB: Yes, you would get very wide agreement on that.

JK: Then our question is can that material process in the brain bring about a change in itself?

DB: Yes, that is the question.

JK: In itself. And if that material in itself can change, it would still be a material process. Right?

DB: Yes. Thought is always apparently going to be a material process.

JK: And therefore it is not insight. We must come back to that.

DB: Right. Now, you are saying that insight is not a material process.

JK: Go slowly. We must be careful to use the right words. Thought is a material process in the brain; and any other movement, springing from that material process, is still material process.

DB: Yes, it has to be.

JK: Right. Is there another activity which is not a material process?

DB: Of course people have asked that question for ages. Is there spirit beyond matter?

JK: Spirit, Holy Ghost! Is there some other activity of the brain which cannot be related to the material process?

DB: Well, it cannot depend upon it. Insight cannot depend on the material process, as it would then be just another material process.

JK: Insight cannot depend on the material process, which is thought.

DB: But you were putting it the other way round, that the material process may depend on insight, may be changed by insight.

JK: Ah, wait. The material process is dependent on it, but insight is not dependent on that process.

DB: Now, many people would not see how something non-material would affect something material.

JK: Yes, quite.

DB: It might be easily agreed that something non-material is not affected by matter, but then how does the operation work the other way?

JK: What do you say? The brain, thought, with its content, is a material process. Any activity from it is still part of that. Now, is insight part of that too?

DB: We have agreed on its independence of that; it can't be part of it. But it can still act within the material process. That's the crucial thing.

JK: Yes. That's right. Insight is independent of the material process, but yet it can act upon it.

DB: Let's discuss that a little. Generally speaking, in science, if A can act on B, there is usually reciprocal action of B on A. We don't find situations where A acts on B, but B never acts on A.

JK: I see, I see.

DB: This is one of the difficulties you have raised. We don't find this elsewhere; in human relations, if I can act on you, you can act on me, right?

JK: Yes, we see that human relationships are interaction.

DB: Yes, mutual relationships.

JK: And in those relationships there is response, and so on. Now, if I don't respond to your action, I am independent of it.

DB: But, you see, science generally finds that it is not possible to have a one-sided action.

JK: Quite. So we are continually insisting that the material process must have a relationship to the other.

DB: An action anyway. "Relationship" is an ambiguous word here. If you said "action" it would be more clear.

JK: All right. The material process must be able to act on the non-material, and the non-material must act on the material.

DB: But that would make them both the same.

JK: Exactly!

Q: Not necessarily. One could envisage that insight is a much larger movement than the material process of the brain, and therefore that the larger movement can act on the smaller movement, but the smaller cannot act on the larger.

JK: Yes, we are saying the same thing.

DB: The small movement has no significant action on the larger movement. You can have a situation that if you drop a rock in the ocean, the ocean absorbs it with no significant change.

JK: Yes.

Q: So then they would still have a two-way action but only one action would be significant.

JK: No, no. Don't enter into that too quickly, let us be careful. Love has no relationship to hate.

DB: Again there is this word "relationship." Would you, for example, say that hate has no action on love?

JK: They are independent.

DB: Independent, they have no action on each other.

JK: Ah, it is a very important thing to discover this. Love is independent of hate. Where there is hate the other cannot exist.

DB: Yes, they can't stand side by side, acting on each other.

JK: They can't. So when scientists say if A has a relationship to B, then B must have a relationship to A, we are contradicting that.

DB: Not all scientists have said that; a few have said otherwise. I don't like to bring in Aristotle . . .

JK: Bring him in!

DB: He said there is an unmoved mover, that God is never moved by matter; he is not acted on by matter, but he acts. Do you see? That is an old idea, then. Since Aristotle's time, science has thrown out this concept and said that it is impossible.

JK: If I see clearly that love is independent of hate, hate cannot possibly act on love. Love may act on hate, but where hate is, the other cannot be.

DB: Well, those are two possibilities. Which are you saying?

JK: What are the two possibilities?

DB: You said one possibility is that love may act on hate, and the other is that they have no action at all on each other.

JK: Yes.

DB: But which?

JK: I understand. No, love cannot act on hate.

DB: Right. They have no relationship. But perhaps insight could, you see.

JK: We have to be quite clear on this point. Violence and being without violence are two entirely different factors. One cannot act upon the other.

DB: In that case you could say that the existence of the one is the nonexistence of the other, and there is no way in which they can act together.

JK: That's right.

DB: They cannot be there together.

JK: Absolutely. I'll stick to that. So when this material process is in action, the other cannot exist.

DB: What is "the other" this time? Insight?

JK: Yes.

DB: That denies what we were saying before; that there is an action from insight on the material process.

JK: Now, steady, yes. Where there is violence, the other—I hate to use the word "nonviolence"—is not.

DB: Peace or harmony?

JK: Where there is violence, peace cannot exist. But where there is peace, is there violence? No, of course not. So peace is independent of violence.

Q: You have said many, many times that intelligence can act upon thought; insight can affect thought, but it doesn't work the other way round. You have given many examples of this.

JK: Intelligence can wipe away ignorance, but ignorance cannot touch intelligence, right? Where there is love, hate can never exist. Can love wipe away hate?

DB: We said that this doesn't seem to be possible, because hate appears to be an independent force.

JK: Of course it is.

DB: It has its own momentum, you see, its own force, its own movement.

Q: I don't quite get this relationship of love and hate with the earlier discussion of insight.

DB: There seem to be two different areas.

Q: Thought is a movement, and insight seems to be non-movement, where everything seemingly is at rest, and it can observe movement.

DB: That is what we are trying to get at, the notion of something which is not affected by anything else.

Q: Aren't you then saying, in looking at love and hate, that there is good and there is evil, and that evil is a completely separate, independent force?

DB: Well, it is independent of good.

Q: But is the process in the mind, or is it related to insight?

DB: We are coming to that.

Q: Take light and darkness. Light appears, and the darkness is gone.

DB: Good and evil; love and hate; light and darkness—when one is, the other can't be, you see. That is all we are saying so far.

Q: Do you mean in a single brain?

DB: In any brain, yes, or in any group or anywhere. Whenever there is hate going on in a group, there is not love.

JK: Something has just come to my mind. Love has no cause. Hate has a cause. Insight has no cause. The material process, as thought, has a cause. Right?

DB: Yes, it is part of the chain of cause and effect.

JK: Can that which has no cause ever act upon that which has a cause?

DB: It might. We can see no reason why that which has no cause might not act on something that has a cause. There is no obvious reason. It won't happen the other way round. What has a cause cannot act on that which has no cause, because that would invalidate it.

JK: That's right. But apparently the action of insight has an extraordinary effect on the material process.

DB: It may, for example, wipe out some causes.

JK: As insight is causeless, it has a definite effect on that which has cause.

DB: Well, it doesn't necessarily follow, but it is possible.

JK: No, no, I don't say it is possible.

DB: I am saying we haven't quite seen why it is necessary. There is no contradiction when we say the word "possible."

JK: All right, I see. As long as we are clear on the word "possible." We must be careful. Love is without cause, and hate has a cause. The two cannot coexist.

DB: Yes. That is true. That is why there is a difference between love and insight. That is why it doesn't follow necessarily that if something has no cause, it will act on something that has a cause. That is what I was trying to say.

JK: I just want to explore a little more. Is love insight?

DB: As far as we can see, it is not the same. Love and insight are not identical, are they? Not exactly the same thing.

JK: Why?

DB: Insight may be love, but, you see, insight also occurs in a flash.

JK: It is a flash of course. And that flash alters the whole pattern, operates on it, uses the pattern, in the sense argues, reasons, uses logic, and all that. I don't know if I am making myself clear.

DB: I think that once the flash has operated, the pattern is different, and would therefore be more rational. The flash may make logic possible, because you may have been confused before the flash.

JK: Yes, yes! Aristotle may have come to all this by logic.

DB: Well, he may have had some insight! We don't know.

JK: We don't know, but I am questioning it.

DB: We really don't know how his mind operated because there are only a few books that have survived.

JK: Would you say by reading some of those books that he had insight?

DB: I haven't really read Aristotle directly; very few people have because it is hard. Most people read what other people have said about Aristotle. A few phrases of his are common, like "the unmoved mover." And he has said some things which suggest that he was quite intelligent, at least.

JK: What I am trying to say is that insight is never partial; I am talking of total, not partial, insight.

Q: Krishnaji, could you explain that a little? What do you mean by "not partial" insight?

JK: An artist can have a partial insight. A scientist can have a partial insight. But we are talking about total insight.

Q: You see the artist is also a human being, so . . .

JK: But his capture of insight is partial.

Q: It is directed to some form of art. So you mean that it illuminates a limited area or subject. Is that what you mean by partial insight?

JK: Yes.

Q: Then what would be total insight? What would it encompass?

JK: The total human activity.

DB: That is one point. But earlier on, we were asking whether this insight would illuminate the brain, the activity of the brain. In that illumination, it seems that the material activity of the brain will change. Would that be correct? We must get this point clear, then we can raise the question of totality. Are we saying that insight is an energy which illuminates the activity of the brain? And that in this illumination, the brain itself begins to act differently.

JK: You are quite right. That's all. That is what takes place. Yes.

DB: We say the source of this illumination is not in the material process; it has no cause.

JK: No cause.

DB: But it is a real energy.

JK: It is pure energy. Is there action without cause?

DB: Yes, without time. Cause implies time.

JK: That is, this flash has altered completely the pattern which the material process has set.

DB: Could you say that the material process generally operates in a kind of darkness, and therefore it has set itself on a wrong path?

JK: In darkness, yes. That is clear. The material process acts in ignorance, in darkness. And this flash of insight enlightens the whole field, which means that ignorance and darkness have been dispelled. I will hold to that.

DB: You could say, then, that darkness and light cannot coexist for obvious reasons. Nevertheless the very existence of light is to change the process of darkness.

JK: Quite right.

Q: But what contributes the flash?

JK: We haven't come to that yet. I want to go step by step into this. What has happened is that the material process has worked in darkness and has brought about confusion and all the mess that exists in the world. But this flash of insight wipes away the darkness. Which means that the material process is not then working in darkness.

DB: Right. But now let's make another point clear. When the flash has gone, the light continues.

JK: The light is there; the flash is the light.

DB: At a certain moment the flash is immediate, but then, as you work from there, there is still light.

JK: Why do you differentiate flash from light?

DB: Simply because the word "flash" suggests something that happens in one moment.

JK: Yes.

DB: You see, we are saying that insight would only last in that moment.

JK: We must go slowly.

DB: Well, it is a matter of language.

JK: Is it merely a matter of language?

DB: Perhaps not, but if you use the word "flash," there is the analogy of lightning, giving light for a moment, but then the next moment you are in darkness, until there is a further flash of lightning.

JK: It is not like that.

DB: So what is it? Is it that the light suddenly turns on and stays on?

JK: No. Because when we say "stays on" or "goes off," we are thinking in terms of time.

DB: We have to clear this up, because it is the question everybody will put.

JK: The material process is working in darkness, in time, in knowledge, in ignorance, and so on. When insight takes place, there is the dispelling of that darkness. That is all we are saying. Insight dispels that darkness. And thought, which is the material process, no longer works in darkness. Therefore that light has altered—no, it has ended—ignorance.

DB: So we say that this darkness is really something which is built into the content of thought.

JK: The content is darkness.

DB: That's right. Then that light has dispelled that ignorance.

JK: That's right. Dispelled the content.

DB: But still we have to be very careful, since we still have content in the usually accepted sense of the word; we know all kinds of things.

JK: Of course.

DB: So we can't say that the light has dispelled *all* the content.

JK: It has dispelled the centre of darkness.

DB: Yes, the source, the creator of darkness.

JK: Which is the self, right? It has dispelled the centre of darkness, which is the self.

DB: We could say that the self, which is part of the content—that part of the content which is the centre of darkness, which creates it and maintains it—is dispelled.

JK: Yes, I hold to that.

DB: We see now that this means a physical change in the brain cells. That centre, that content which is the centre, is a certain set, form, disposition, of all the brain cells, and it in some way alters.

JK: Obviously! You see, this has enormous significance, in our relationship with our society, in everything. Now, the next question is how does this flash come about? Let's begin the other way round. How does love come about? How does peace come about?

Peace is causeless; violence has cause. How does that causeless thing come about when my whole life is causation? No, there is no "how," right? The "how" implies a cause, so there is no "how."

Q: Are you saying that since it is without cause, it is something that just exists?

JK: No, I don't say that it exists. That is a dangerous statement.

Q: It has to exist at some point.

JK: No. The moment you say it exists, it is not.

DB: You see, the danger is that it is part of the content.

JK: The question you put was about a mutation in the brain calls. That question has been put after a series of discussions. And we have come to a point when we say that the flash, that light, has no cause; that the light operates on that which has cause, which is the darkness. That darkness exists as long as the self is there; it is the originator of that darkness, but light dispels the very centre of darkness. That's all. We have come to that point. And therefore there is a mutation. Then I say that the question of how do I get this flash of insight, how does it happen, is a wrong question. There is no "how."

Q: There is no "how," but there is darkness and there is light.

JK: Just see first there is no "how." If you show me how, you are back into the darkness. Right?

DB: Yes.

JK: It is a tremendous thing to understand that. I am asking something else, which is why is it that we have no insight at all? Why is it that this insight doesn't start from our childhood?

DB: Well, the way life is lived . . .

JK: No, I want to find out. Is it because of our education? Our society? I don't believe it is all that. You follow?

DB: What do you say, then?

JK: Is it some other factor? I am groping after this. Why don't we have it? It seems so natural.

DB: At first, one would say something is interfering with it.

JK: But it seems so natural. For X, it is quite natural. Why isn't it natural for everyone? Why isn't it possible? If we talk about blockages, education, and so on, which are all in the realm of causation, then to remove the blockages implies another cause. So we keep on rolling in that direction. There is something unnatural about all this.

Q: If you would say that there are blocks . . .

JK: I don't want to use that; it is the language of the darkness.

Q: Then you could say that the blocks prevent the insight from acting.

JK: Of course. But I want to move away from these blockages.

DB: Not exactly blockages, but we used the words "centre of darkness," which we say is maintaining darkness.

JK: Why isn't it natural for everybody to have this insight?

DB: That is the question.

JK: Why is love not natural to everybody? Am I putting the question clearly?

DB: I think, to make it more clear, some people might feel that it is natural to everybody, but being treated in a certain way, they gradually get caught in hate.

JK: I don't believe that.

DB: Then you would have to suppose that the young child meeting hate would not respond with hate.

JK: Yes, that's right.

DB: Most people would say that it is natural for the young child meeting hate to respond with hate.

JK: Yes, this morning I heard that. Then I asked myself why?. Now, just a minute. X has been put under all these circumstances, which could have produced blockages, but X wasn't touched by them. So why is it not possible for everybody?

DB: We should make it clear why we say it would be natural not to respond to hate with hate.

JK: All right. Limit it to that.

DB: Even when one hasn't thought about it. You know, the child is not able to think about all this. Some people would say it is instinct, the animal instinct . . .

JK: . . . which is to hate.

DB: Well, to fight back.

JK: To fight back.

DB: The animal will respond with love, if you treat him with love, but if you treat the animal with hate he is going to fight back.

JK: Of course.

DB: He will become very vicious.

JK: Yes.

DB: Now, some people would say that the human being in the beginning is like that animal, and later he can understand.

JK: Of course. That is, the human being's origins were with the animal, and the animal, the ape or the wolf . . .

DB: The wolf will respond with love too.

JK: And we are saying, why . . .

DB: Look, almost everybody feels that what I said is true, that when we are very young children, we are like the animal. Now you are asking, why don't all young children immediately fail to respond to hate with hate?

JK: That means is it the fault of the parents?

DB: What you are implying is that it is not entirely that. There must be something deeper.

JK: Yes, I think there is something quite different. I want to capture that.

DB: This is something that would be important.

JK: How do we find out? Let's have an insight! I feel that there is something totally different. We are attacking it from a causational point of view. Would it be right to say that the beginning of man is not animal?

DB: Well, that is not clear. The present theory of evolution is that there have been apes, developing; you can follow the line where they become more and more like human beings. Now, when you say that the beginning of man is not animal, it is not clear.

JK: If the beginning of man is the animal, therefore that instinct is natural and then it is highly cultivated.

DB: Yes, that instinct is cause and effect.

JK: Cause and effect, and it becomes natural. But someone comes along and asks, "Is it?"

DB: Let's try to get this clear.

JK: I mean scientists and historians have said that man began from the ape, and that, as all animals respond to love and to hate, we as human beings respond instantly to hate by hate.

DB: And vice versa, to love by love.

JK: At the beginning there were people, half a dozen people who never responded to hate, because they had love. And of those few, one or two also implanted this thing in the human mind, right? That where love is, hate is not. And that has also been part of our inheritance. Why have we cultivated the response of hate to hate? Why haven't we cultivated the other? Oh, the other—love—is something that cannot be cultivated.

DB: It is not causal. Cultivation depends on a cause.

JK: On thought. So why have we lost the other? We have cultivated very carefully, by thought, the concept of meeting hate by hate, violence by violence, and so on. Why haven't we moved along with the other line? With love, that is causeless? You follow my question?

DB: Yes.

JK: Is this a futile question?

DB: One doesn't see any way of proceeding.

JK: I am not trying to proceed.

DB: We have to understand what made people respond to hate with hate . . .

JK: To X, the other seems so natural. So if that is so natural to him, why isn't it natural to everyone else? It must be natural to others! You know this ancient idea, which is probably in existence in the Jewish and in the Indian religions and so on, that the manifestation of the highest takes place, occasionally. That seems too easy an explanation. Has mankind moved in the wrong direction? Have we taken a wrong turn?

DB: Yes, we have discussed this before, that there has been a wrong turning.

JK: To respond to hate by hate, violence by violence, and so on.

DB: And to give supreme value to knowledge.

Q: Wouldn't another factor also be the attempt to cultivate the idea of love? The purpose of the religions has been to produce love and better human beings.

JK: Don't go into all that. Love has no cause; it is not cultivatable. Full stop.

Q: Yes, but the mind doesn't see that.

JK: But we have explained all that. I want to find out why, if it is natural to X, it isn't natural to others. I think this is a valid question.

DB: Another point is to say that you could see that the response of hate to hate makes no sense anyway. So why do we go on with it? Because many believe in that moment that they are protecting themselves with hate, but it is no protection.

JK: But to go back to that question, I think it is valid. A, B, C are born without cause; X, Y, Z are caught in cause. Why? You understand? Is it the privilege of the few? The elite? No, no. Let's look at it another way. X, Y, Z's minds are the mind of humanity; we've been through that. The mind of humanity has been responding to hate with hate, violence by violence, and knowledge by knowledge. And A, B, C are part of humanity and do not respond to hate by hate, like X, Y, and Z! They are part of X, Y and Z's consciousness, part of all that.

DB: Why is there this difference?

JK: That is what I am asking. One is natural, the other is unnatural. Why? Why the difference? Who is asking this question? X, Y, and Z, who respond to hate by hate, are they asking the question? Or are A, B, and C asking the question?

Q: It would seem that A, B, and C are asking it.

DB: Yes, but you see we were also just saying that they are not different. We say they *are* different, but also that they are *not* different.

JK: Of course. They are not different.

DB: There is one mind.

JK: That's it, one mind.

DB: Yes, and how does it come that another part of this one mind says no?

JK: That's the whole thing. How does it come about that one part of the mind says we are different from another? Of course, there are all kinds of explanations, and I am left with the fact that A, B, and C are different from X, Y, and Z. And those are facts, right?

Q: They appear to be different.

JK: Oh, no.

Q: They are actually different.

JK: Absolutely, not just apparently.

DB: I think the question we want to come back to is why do the people who cultivate hate say that they are different from those who don't?

JK: Do they say that?

DB: I think they do, insofar as they would admit that if there was anybody who didn't cultivate hate, they must be different.

JK: Yes, that is clear—light and darkness, and so on. But I want to find out if we are moving in the right direction. That is, A, B, and C have given me that gift, and I have not carried that gift. You follow what I mean? I have carried the other. Why? If a father has responded to hate by hate, why has the son not responded in the same way?

DB: I think it is a question of insight.

JK: Which means that the son had insight right from the beginning. You follow what I am saying? Right from childhood, which means what?

DB: What?

JK: I don't want to enter into this dangerous field yet!

DB: What is it? Perhaps you want to leave that.

JK: There is some factor that is missing. I want to capture it. You see, if that is an exception, then it is silly.

DB: All right. Then we agree that the thing is dormant in all human beings. Is that what you want to say?

JK: I am not quite sure that is what I want to say.

DB: But I meant that the factor is there in all mankind.

JK: That is a dangerous statement too.

DB: That is what you were saying.

JK: I know, but I am questioning. When I am quite sure, I will tell you [*laughs*].

DB: All right. We tried this, and we can say it seems promising but it is a bit dangerous. This possibility is there in all mankind, and insofar as some people have seen it.

JK: Which means God is in you?

DB: No, it is just that the possibility of insight is there.

JK: Yes, partly. I am questioning all this. The father responds to hate by hate; the son doesn't.

DB: That happens from time to time.

JK: No, consistently from the beginning. Why?

DB: It must depend on insight, which shows the futility of hate.

JK: Why did that man have it?

DB: Yes, why?

JK: And why if this seems so terribly natural to him, is it not natural to everybody? As water is natural to everybody.

DB: Well, why isn't insight present for everybody from the beginning?

JK: Yes, that is what I am asking.

DB: So strongly that even maltreatment cannot affect it.

JK: Nothing can affect it. That is my point. Maltreatment, beating, being put into all kinds of dreadful situations hasn't affected it. Why? We are coming to something.

Seven

Death Has Very Little Meaning

JIDDU KRISHNAMURTI: Shall we start where we left off? Are we saying that human beings are still behaving with the animal instincts?

DAVID BOHM: Yes, and that the animal instincts, it seems, may be overpowering in their intensity and speed, and especially with young children. It may be that it is only natural for them to respond with the animal instinct.

JK: So that means, after a million years, that we are still instinctively behaving like our ancestors?

DB: In some ways. Probably our behaviour is also complicated by thought; the animal instinct has now become entangled with thought, and it is getting in some ways worse.

JK: Far worse.

DB: Because all these instincts of hatred now become directed and sustained by thought, so that they are more subtle and dangerous.

JK: And during all these many centuries we haven't found a way, a method, a system—something that will move us away from that track. Is that it?

DB: Yes. One of the difficulties, surely, is that when people begin to be angry with each other, their anger builds up and they can't seem to do anything about it. They may try to control it, but that doesn't work.

JK: As we were saying, someone—X—behaves naturally in a way that is not a response to the animal instinct. What place has such insight in human society? None at all?

DB: In society as it is, it cannot be accommodated, because society is organized under the assumption that pain and pleasure and fear are going to rule, except when you control them. You could say that friendliness is a kind of animal instinct too, for people become friendly for instinctive reasons. And perhaps they become enemies for similar reasons.

So I think that some people would say that we should be rational rather than instinctive. There was a period during the eighteenth century, the Age of Reason, when they said man could be rational, could choose to be rational, in order to bring about harmony everywhere.

JK: But he hasn't done so!

DB: No, things got worse, leading to the French Revolution, to the Terror, and so on. So after that, people didn't have so much faith in reason as a way of getting anywhere or coming out of conflict.

JK: So where does that lead us? We were talking really about insight that actually changes the nature of the brain itself.

DB: Yes, by dispelling the darkness in the brain, insight allows the brain to function in a new way.

JK: Thought has been operating in darkness, creating its own darkness and functioning in that. And insight is, as we said, like a flash which breaks down the darkness. Then, when that insight clears the darkness, does man act, or function, rationally?

DB: Yes, man will then function rationally and through perception, rather than just by rules and reason. But there is a freely flowing reason. You see, some people identify reason with certain rules of logic which would be mechanical. But there can be reason as a form of perception of order.

JK: So we are saying, are we, that insight is perception?

DB: It is the flash of light which makes perception possible.

JK: Right. That's it.

DB: It is even more fundamental than perception.

JK: So insight is pure perception, and from that perception there is action, which is then sustained by rationality. Is that it?

DB: Yes.

JK: That's right.

DB: And the rationality is perception of order.

JK: So would you say there is insight, perception, and order?

DB: Yes.

JK: But that order is not mechanical because it is not based on logic.

DB: There are no rules.

JK: No rules. Let's put it that way: It's better. This order is not based on rules. This means insight, perception, action, order. Then you come to the question, is insight continuous, or is it by flashes?

DB: We went into that and felt it was a wrong question, so perhaps we can look at it differently. It is not time-binding.

JK: Not time-binding. Yes, we agreed on that. So now let's get a little further. We said, didn't we, that insight is the elimination of the darkness which is the very centre of the self, the darkness that self creates. Insight dispels that very centre.

DB: Yes. With the darkness, perception is not possible. It's blindness in a way.

JK: Right. Then what next? I am an ordinary man, with all my animal instincts, pleasure and pain and reward and punishment and so on. I hear you say this, and I see what you are saying has some kind of reason, logic, and order.

DB: Yes, it makes sense as far as we can see it.

JK: It makes sense. Then how am I to have it in my daily life? How am I to bring it about? You understand that these words, which are difficult, are all of them time-binding. But is that possible?

DB: Yes, without time, you see.

JK: Is it possible for man, with his narrow mind, to have this insight, so that that pattern of life is broken? As we said the other day, we have tried all this, tried every form of self-denial, and yet that insight doesn't come about. Once in a while there is a partial insight, but that partial insight is not the whole insight, so there is still partial darkness.

DB: Which doesn't dispel the centre of the self. It may dispel some darkness in a certain area, but the source of the darkness, the creator, the sustainer of it, is still there.

JK: Still there. Now, what shall we do? But this is a wrong question. This leads nowhere. We have stated the general plan, right? And I have to make the moves, or make no moves at all. I haven't the energy. I haven't the capacity to see it quickly. Because this is immediate, not just something that I practise and eventually get. I haven't the capacity, I haven't the sense of urgency, of immediacy. Everything is against me: my family, my wife, society. Everything. And does this mean that I eventually have to become a monk?

DB: No. Becoming a monk is the same as becoming anything else.

JK: That's right. Becoming a monk is like becoming a businessman! I see all this, verbally as well as rationally, intellectually, but I can't capture this thing. Is there a different approach to this problem? I am always asking the same question, because I am caught in the same pattern. So is there a totally different way? A totally different approach to the whole turmoil of life? Is there a different manner of looking at it? Or is the old way the only way?

We have said that as long as the centre is creating darkness, and thought is operating in that darkness, there must be disorder, and society will be as it is now. To move away from that, you must have insight. Insight can only come about when there is a flash, a sudden light, which abolishes not only darkness but the creator of darkness.

DB: Yes.

JK: Now I am asking if there is a different approach to this question altogether, although the old response seems so absolute.

DB: Well, possibly. When you say it seems absolute, do you want a less absolute approach?

JK: I am saying that if that is the only way, then we are doomed.

DB: You can't produce this flash at will.

JK: No, it can't be produced through will, through sacrifice, through any form of human effort. That is out; we know we have finished with all that. And also we agreed that to some people—to X—this insight seemed so natural, and we asked why is it not natural to others.

DB: If we begin with the child, it seems natural to the child to respond with his animal instincts, with great intensity which sweeps him away. Darkness arises because it is so overwhelming.

JK: Yes, but why is it different with X?

DB: First of all it seems natural to most people that the animal instincts would take over.

JK: Yes, that's right.

DB: And they would say the other fellow, X, is unnatural.

JK: Yes.

DB: So that is the way mankind has been thinking, saying that if there are indeed any people who are different, they must be very unusual and unnatural.

JK: That's it. Human beings have been responding to hatred by hatred and so on. There are those few, perhaps many, who say that is not natural or rational. Why has this division taken place?

DB: If we say that pleasure and pain, fear and hate, are natural, then it is felt that we must battle to control these; otherwise they will destroy us. The best we can hope for is to control them with reason, or through another way.

JK: But that doesn't work! Are people like X, who function differently, the privileged few, by some miracle, by some strange chance event?

DB: Many people would say that.

JK: But it goes against one's grain. I would not accept that.

DB: Well, if that is not the case, then you have to say why there is this difference.

JK: That is what I am trying to get at, because X is born of the same parents.

DB: Yes, fundamentally the same, so why does he behave differently?

JK: This question has been asked many times, over and over again in different parts of the world. Now, why is there this division?

QUESTIONER: Is the division really total? You see, even the man who responds to hatred with hatred nevertheless sees that it doesn't make sense, is not natural, and should be different.

JK: It should be different, but he is still battling with ideas. He is trying to get out of it by the exercise of thought which breeds darkness.

Q: I just want to say that the division does not seem to be so entire.

JK: Oh, but the division is entire, complete.

Q: Well, then, why are people not simply saying, let's continue to live that way, kill each other, and let's enjoy it to the last moment?

JK: Because they can't see anything except their own darkness.

Q: But they want to get out of it.

JK: Now, wait a minute. Do they want to get out of it? Do they actually realize the state they are in and deliberately want to get out of it?

Q: They are ambivalent about it. They want to go on getting the fruits of it, but they have a sense that it is wrong and that it leads to suffering.

DB: Or else they find they can't help it. You see, when the time comes to experience anger or pleasure, they can't get away.

JK: They can't help it.

Q: But they want to get out of it, although they are helpless. There are forces which are stronger than their will.

JK: So what shall we do? Or is this division false?

DB: That's the point. We had better talk of a difference between these two approaches. This difference is not fundamental.

JK: I don't think they have anything in common.

DB: Why? You say the difference is false, although fundamentally people are the same, but a difference has developed between them. Perhaps most people have taken a wrong turning.

JK: Yes, let's put it that way.

DB: But the difference is not intrinsic, it is not structural, built in like the difference between a tree and a rock.

JK: Agreed. As you say, there is a difference between a rock and a tree, but it is not like that. Let's be simple. There are two responses. They start from the source. One has taken one direction, and the other has taken a different direction. But the source is the same. Why haven't all of them moved in the right direction?

DB: We haven't managed to answer that. I was just saying that if one understands that, then going back to the source, one does not have to take the wrong turn. In a sense we are continually taking this wrong turn, so if we can understand this, then it becomes possible to change. And we are continually starting from the same source, not going back in time to a source.

JK: Just a minute, just a minute.

DB: There are two possible ways of taking your statement. One is to say that the source is in time, that far back in the past we started together and took different paths. The other is to say that the source is timeless, and we are continually taking the wrong turn, again and again. Right?

JK: Yes, it is constantly the wrong turn. Why?

Q: This means that there is the constant possibility of the right turn.

JK: Yes, of course. That's it. If we say there is a source from which we all began, then we are caught in time.

DB: We can't go back.

JK: No, that is out. Therefore it is apparent that we are taking the wrong turn all the time.

DB: Constantly.

JK: Constantly taking the wrong turn. But why? The one who is living with insight and the other who is not living with insight—are these constant? The man who is living in darkness can move away at any time to the other. That is the point. At any time.

DB: Then nothing holds him, except constantly taking the wrong turn. You could say the darkness is such that he doesn't see himself taking the wrong turn.

JK: Are we pursuing the right direction, putting the right question? Suppose you have that insight, and your darkness, the very centre of darkness, has been dispelled completely. And I, a serious, fairly intelligent, not neurotic human being, listen to you. And whatever you have said seems reasonable, rational, sane. I question the division. The division is created by the centre which creates darkness. Thought has created it.

DB: Well, in darkness, thought creates the division. From the darkness a shadow is thrown; it makes a division.

JK: If we have that insight, we say there is no division, and man won't accept that, because in his darkness there is nothing but division. So we, living in darkness, have created the division. We have created it in our thoughts...

DB: We are constantly creating it.

JK: Yes, always wanting to live constantly in a state in which there is no division. That movement, however, is still the movement of darkness. Right?

DB: Yes.

JK: How am I to dispel this continuous, constant darkness? That is the only question, because as long as that exists, I create this constant division. You see, this is going round in circles. I can only dispel the darkness through insight, and I cannot have that insight by any effort of will, so I am left with nothing. So what is my problem? My problem is to perceive the darkness, to perceive the thought that is creating darkness, and to see that the self is the source of this darkness. Why can't I see that? Why can't I see it even logically?

DB: Well, it's clear logically.

JK: Yes, but somehow it doesn't seem to operate. So what shall I do? I realize for the first time that the self is creating the darkness which is constantly breeding division. I see that very clearly.

DB: And the division produces the darkness anyway.

JK: Vice versa, back and forth. And from all that, everything begins. I see that very clearly. What shall I do? So I don't admit division.

Q: Krishnaji, aren't we introducing division again, nevertheless, when we say there is the man who needs insight?

JK: But man, X, has insight, and he has explained very clearly how darkness has vanished. I listen to him, and he says your very darkness is creating the division. Actually, there is no division, no division as light and darkness. So he asks me, "Can you banish, can you put away this sense of division?"

DB: You seem to be bringing back a division by saying that, by saying that I should do it, you see.

JK: No, not "should."

DB: In a way you are saying that the thought process of the mind seems spontaneously to produce division. You say, try to put it aside, and at the same time it is trying to make division.

JK: I understand. But can my mind put away division? Or is that a wrong question?

Q: Can it put away division as long as it is divided?

JK: No, it can't. So what am I to do? Listen, it is not division. X says something so extraordinarily true, of such immense significance and beauty that my whole being says, "Capture it." That is not a division.

I recognize that I am the creator of division, because I am living in darkness, and so out of that darkness I create. But I have listened to X, who says there is no division. And I recognize that is an extraordinary statement. So the very saying of that to one who has lived in constant division has an immediate effect. Right?

DB: I think that one has to, as you say, put away the division ...

JK: I will leave that; I won't put away. That statement that there is no division—I want to get at that a little bit. I am getting somewhere with it. X's statement from this insight, that there is no division, has a tremendous effect on me. I have lived constantly in division, and he comes along, after discussing it, and says there is no division. What effect has it on me? It must have some effect on me, otherwise what is the good of talking?

DB: Then you say there is no division. That makes sense. But on the other hand it seems that the division exists.

JK: I recognize the division, but the statement that there is no division has this immense impact on me. That seems natural, doesn't it? When I see something that is immovable, it must have some effect on me. I respond to it with a tremendous shock.

DB: You see, if you were talking about something which was in front of us, and you said, "No, it is not that way," then that would, of course, change your whole way of seeing it. Now, you say this division is not that way. We try to look and see if that is so, right?

JK: I don't even say, "Is that so?" X has very carefully explained the whole business, and he says at the end of it that there is no division. And I am sensitive, watching very carefully, and realizing that I am constantly living in division. When X makes that statement, it has broken the pattern.

I don't know if you follow what I am trying to explain. It has broken the pattern, because he has said something which is so fundamentally true. There is no God and man. Right, sir. I stick to that. I see something—which is where hatred exists the other is not. But, hating, I want the other. So constant division is born out of darkness. And the darkness is constant. But I have been listening very carefully, and X makes a statement which seems absolutely true. That enters into me, and the act of his statement dispels the darkness. I am not making an effort to get rid of darkness, but X is the light. That's right, I hold to that.

So it comes to something, which is can I listen in my darkness, which is constant? In that darkness, can I listen to you? Of course I can. I am living in constant division, which brings darkness. X comes along and tells me there is no division.

DB: Right. Now, why do you say you can listen in the darkness?

JK: Oh, yes, I can listen in darkness. If I can't, I am doomed.

DB: But that is no argument.

JK: Of course that is no argument, but it is so!

DB: Living in darkness is not worthwhile. But now we say that it is possible to listen in the darkness.

JK: He, X, explains to me very, very carefully. And I am sensitive. I have been listening to him in my darkness, but that is making me sensitive, alive, watching. That is what I have been doing. We have been doing it together. And he makes a statement that there is absolutely no division. And I know that I am living in division. That very statement has brought the constant movement to an end. Otherwise, if this doesn't take place, I have nothing. You follow? I am perpetually living in darkness. But there is a voice in the wilderness [*laughs*] and listening to that voice has an extraordinary effect.

DB: Listening reaches the source of the movement, whereas observation does not.

JK: Yes, I have observed, I have listened, I have played all kinds of games all my life. And I now see that there is only one thing. That there is this constant darkness and I am acting in the darkness, in this wilderness which is darkness, whose centre is the self. I see that *absolutely*, completely; I can't argue against it anymore. And X comes along and tells me this. In that wilderness a voice says there is water. You follow? It is not hope. There is immediate action in me.

One must realize that this constant movement in darkness is my life. Would I admit that, sir? You follow what I am saying? Can I, with all the experience, with all the knowledge which I have gathered over a million years, suddenly realize that I am living in total darkness? Nobody will admit that. Because that means I have reached the end of all hope, right? But my hope is also darkness. The future is out altogether, so I am left with this enormous darkness, and I am there. That means the realisation of that is the ending of becoming. I have reached that point and X tells me, naturally, sir . . .

You see, all the religions have said that the division exists. God and son of God.

DB: But they say it can be overcome.

JK: It is the same pattern repeated. It doesn't matter who said it, but the fact is somebody in this wilderness is saying something, and in that wilderness I have been listening to every voice, and to my own voice, which has created more and more darkness. Yet this is right. That means, doesn't it, that when there is insight, there is no division.

DB: Yes.

JK: It is not your insight or my insight; it is insight. In that there is no division.

DB: Yes.

JK: Which brings us to that ground we spoke of . . .

DB: What about the ground?

JK: In that ground, there is no darkness as darkness or light as light. In that ground, there is no division. Nothing is born of will or time or thought.

DB: Are you saying that light and darkness are not divided?

JK: Right.

DB: Which means to say there is neither.

JK: Neither—that's it! There is something else. There is a perception that there is a different movement, which is "non-dualistic."

DB: Non-dualistic means what? No division.

JK: No division. I won't use "non-dualistic." There is no division.

DB: But nevertheless there is movement.

JK: Of course.

Q: What does that mean now, without division?

JK: I mean by movement, that movement which is not time. That movement doesn't breed division. So I want to go back, lead to the ground. If, in that ground, there is neither darkness nor light, no God or the son of God—there is no division—what takes place? Would you say that the ground is movement?

DB: Well, it could be, yes. Movement that is undivided.

JK: No, no, no.

DB: You were saying before that there is movement, right?

JK: I say there is movement in darkness.

DB: Yes, but we said, of the ground, there is no division of darkness and light, and you said there is movement.

JK: Yes. Would you say the ground is endless movement?

DB: Yes.

JK: What does that mean?

DB: Well, it is difficult to express.

JK: Keep on going into it; let's express it. What is movement, apart from movement from here to there, apart from time—is there any other movement?

DB: Yes

JK: There is. The movement from being to becoming, psychologically. There is the movement of distance, there is the movement of time. We say those are all divisions. Is there a movement which in itself has no division? When you have made that statement that there is no division, there is that movement surely?

DB: Well, are you saying that when there is no division, that movement is there?

JK: Yes, and I said—X says—that is the ground.

DB: Right.

JK: Would you say it has no end, no beginning?

DB: Yes.

JK: Which means again time.

Q: Can one say that movement has no form?

JK: No form—all that. I want to go a little further. What I am asking is, we said that when you have stated there is no division, this means no division in movement.

DB: It flows without division, you see.

JK: Yes, it is a movement in which there is no division. Do I capture the significance of that? Do I understand the depth of that statement? A movement in which there is no division, which means no time, no distance as we know it. No element of time in it at all. So I am trying to see if that movement is surrounding man?

DB: Yes, enveloping.

JK: I want to get at this. I am concerned with mankind, humanity, which is me. X has made several statements, and I have captured a statement which seems so absolutely true—that there is no division. Which means that there is no action which is divisive.

DB: Yes.

JK: I see that. And I also ask, "Is that movement without time?" It seems that it is the world. You follow?

DB: The universe.

JK: The universe, the cosmos, the whole.

DB: The totality.

JK: Totality. Isn't there a statement in the Jewish world, "Only God can say 'I am'"?

DB: Well, that's the way the language is built. It is not necessary to state it.

JK: No, I understand. You follow what I am trying to get at?

DB: Yes, that only this movement *is*.

JK: Can the mind be of that movement? Because that is timeless, therefore deathless.

DB: Yes, the movement is without death; insofar as the mind takes part in that, it is the same.

JK: You understand what I am saying?

DB: Yes. But what dies when the individual dies?

JK: That has no meaning, because once I have understood there is no division ...

DB: Then it is not important.

JK: Death has no meaning.

DB: It still has a meaning in some other context.

JK: Oh, the ending of the body—that's totally trivial. But you understand? I want to capture the significance of the statement that there is no division. It has broken the spell of my darkness, and I see that there is a movement, and that's all. Which means death has very little meaning.

DB: Yes.

JK: You have abolished totally the fear of death.

DB: Yes, I understand that when the mind is partaking in that movement, then the mind is that movement.

JK: That's all! The mind is that movement.

DB: Would you say that matter is also that movement?

JK: Yes, I would say everything is. In my darkness I have listened to X. That's most important. And his clarity has broken my spell. When he said there is no division, he abolished the division between life and death. I don't know if you see this.

DB: Yes.

JK: One can never say then "I am immortal." It is so childish.

DB: Yes, that's the division.

JK: Or "I am seeking immortality." Or "I am becoming." We have wiped away the whole sense of moving in darkness.

Q: What then would be the significance of the world? Is there a significance to it?

JK: The world?

Q: With man.

DB: Society, do you mean?

Q: Yes, it seems that when you make that statement, there is no division, and life is death—what then is the significance of man with all his struggle?

JK: Man in darkness. What importance has that? It is like struggling in a locked room. That is the whole point.

DB: Significance can only rise when the darkness is dispelled.

JK: Of course.

Q: The only significance is the dispelling of the darkness.

JK: Oh, no, no!

DB: Aren't we going to say that something more can be done besides dispelling the darkness?

JK: I have listened very carefully to everything that you, who have insight, say. What you have done is to dispel the centre. In darkness I could invent many things of significance; that there is light, there is God, there is beauty, there is this and that. But it is still in the area of darkness. Caught in a room full of darkness, I can invent a lot of pictures, but I want to get something else. Is the mind of the one who has this insight—who therefore dispels darkness and has understanding of the ground which is movement without time—is that mind itself that movement?

DB: Yes, but it isn't the totality. The mind is the movement, but we are saying movement is matter, movement is mind. And we were saying that the ground may be beyond the universal mind. You said earlier that the movement, that the ground, is more than the universal mind, more than the emptiness.

JK: We said that. Much more.

DB: Much more. But we have to get this clear. We say that the mind is this movement.

JK: Yes, mind is the movement.

DB: We are not saying that this movement is only mind?

JK: No, no, no.

DB: That is the point I was trying to get correct.

JK: Mind is the movement—mind in the sense "the ground."

DB: But you said that the ground goes beyond the mind.

JK: Now, just a minute. What do you mean by "beyond the mind"?

DB: Just going back to what we were discussing a few days ago. We said we have the emptiness, the universal mind, and then the ground is beyond that.

JK: Would you say beyond that is this movement?

DB: Yes. The mind emerges from the movement as a ground and falls back to the ground; that is what we are saying.

JK: Yes, that's right. Mind emerges from the movement.

DB: And it dies back into the movement.

JK: That's right. It has its being in the movement.

DB: Yes, and matter also.

JK: So what I want to get at is I am a human being faced with this ending and beginning. And X abolishes that.

DB: Yes. It is not fundamental.

JK: It is not fundamental. One of the greatest fears of life, which is death, has been removed.

DB: Yes.

JK: You see what it does to a human being when there is no death? It means the mind doesn't age—the ordinary mind I am talking about. I don't know if I am conveying this.

DB: Let's go slowly. You say the mind does not age, but what if the brain cells age?

JK: I question it.

DB: But how can we know that?

JK: Because there is no conflict, because there is no strain, there is no becoming, no movement.

DB: This is something that it is hard to communicate with certainty about.

JK: Of course. You can't prove any of this.

DB: But the other, what we have said so far . . .

JK: ... can be reasoned.

DB: It is reason, and also you can feel it. But now you are stating something about the brain cells that I have no feeling for. It might be so; it could be so.

JK: I think it is so. I want to discuss it. When a mind has lived in the darkness and is in constant movement, there is the wearing out, the decay of the cells.

DB: We could say that this conflict will cause cells to decay. But somebody might argue that perhaps even without conflict they could decay at a slower rate. Let's say if you were to live hundreds of years, for example, in time the cells would decay no matter what you did.

JK: Go into this slowly.

DB: I can readily accept that the rate of decay of the cells could be cut down when we get rid of conflict.

JK: Decay can be slowed down.

DB: Perhaps a great deal.

JK: A great deal. Ninety percent.

DB: That we could understand. But if you say a hundred percent, then it is hard to understand.

JK: Ninety percent. Wait a minute. It can be very, very greatly slowed down. And that means what? What happens to a mind that has no conflict? What is that mind, what is the quality of that mind which has no problem? You see, suppose such a mind lives in pure unpolluted air, having the right kind of food and so on. Why can't it live two hundred years?

DB: Well, it is possible; some people have lived for a hundred and fifty years, living in very pure air and eating good food.

JK: But, you see, if those very people who have lived a hundred and fifty years had no conflict, they might live very much longer.

DB: They might. There was a case I was reading of a man in England who lived to be a hundred and fifty. And the doctors became interested in him. They wined and dined him, and then he died in a few days!

JK: Poor devil!

Q: Krishnaji, you generally say that anything that lives in time also dies in time.

JK: Yes, but the brain, which has had insight, has changed the cells.

Q: Are you implying that even the organic brain does not live in time anymore?

JK: No, don't bring in time yet. We are saying that insight brings about a change in the brain cells. Which means that the brain cells are no longer thinking in terms of time.

Q: Psychological time?

JK: Of course. That is understood.

DB: If they are not so disturbed, they will remain in order and perhaps they will break down more slowly. We might increase the age limit from one hundred and fifty to two hundred years, provided one also had healthy living at all levels.

JK: Yes, but all that sounds so damn trivial [*laughs*]!

DB: Yes. It doesn't seem to make much difference, although it is an interesting idea.

JK: What if I live another hundred years? We are trying to find out what effect this extraordinary movement has on the brain.

DB: Yes. If we say the brain is in some way directly enveloped in this movement, that would bring it to order, that there is a real direct flow, physically.

JK: Not only physically.

DB: But also mentally.

JK: Yes, both. It must have an extraordinary effect on the brain.

Q: You talked earlier about energy. Not the everyday energy . . .

JK: We said that that movement is total energy. Now, this insight has captured, seen, that extraordinary movement, and it is part of that energy. I want to come much closer to earth. I have lived with the fear of death, fear of not becoming, and so on. Suddenly I see there is no division, and I understand the whole thing. So what has happened to my brain. You follow?

Let's *see* something. See this whole thing, not verbally but as a tremendous reality, as truth. With all your heart, mind, you see this thing. That very perception must affect your brain.

DB: Yes. It brings order.

JK: Not only order in life but in the brain.

DB: People can prove that if we are under stress, the brain cells start to break down. And if you have order in the brain cells, then it is quite different.

JK: I have a feeling, sir—don't laugh at it; it may be false, it may be true—I feel that the brain never loses the quality of that movement.

DB: Once it has it.

JK: Of course. I am talking of the person who has been through all this.

DB: So probably the brain never loses that quality.

JK: Therefore it is no longer involved in time.

DB: It would no longer be dominated by time. The brain, from what we were saying, is not evolving in any sense; it is just a confusion. You can't say that man's brain has evolved during the last ten thousand years. You see science, knowledge, has evolved, but people felt the same about life several thousand years ago as they do now.

JK: I want to find out. In that silent emptiness that we went through, is the brain absolutely still? In the sense, no movement?

DB: Not absolutely. You see, the blood is going in the brain.

JK: We are not talking of that.

DB: What kind of movement are we discussing?

JK: I am talking of the movement of thought, the movement of any reaction.

DB: Yes. There is no movement in which the brain moves independently. You were saying that there is the movement of the whole, but the brain does not go off on its own, as thought.

JK: You see, you have abolished death, which is a tremendously significant thing. And so I say, what is the brain, the mind, when there is no death? You follow? It has undergone a surgical operation.

DB: Well, the brain normally has the notion of death continually there in the background, and that notion is constantly disturbing the brain, because the brain foresees death, and it is trying to stop it.

JK: To stop the ending of itself and so on.

DB: It foresees all that and thinks it must stop it, but it can't.

JK: It can't.

DB: And therefore it has a problem.

JK: A constant struggle with it. So all that has come to an end. What an extraordinary thing has taken place! How does it affect my daily life? Because I have to live on this earth. My daily life is aggression, this everlasting becoming, striving for success—all that has gone. We will pursue this, but we have understood a great deal today.

DB: In bringing in the question of daily life you might bring in the question of compassion.

JK: Of course. Is that movement compassion?

DB: It would be beyond.

JK: That's it. That's why one must be awfully careful.

DB: Then again, compassion might emerge out of it.

Eight

Can Insight Be
Awakened in Another?

19 APRIL 1980, OJAI, CALIFORNIA

JIDDU KRISHNAMURTI: We were discussing non-movement.
When a human being has been pursuing the path of becoming,
and has gone through all that and this sense of emptiness, silence,
and energy, he has abandoned almost everything and come to the
point, the ground. So how does this insight affect his daily life?
What is his relationship to society? What is his action with regard
to war and the whole world—a world that is really living and
struggling in darkness? What is his action? I would say, as we
agreed the other day, that it is non-movement.

DAVID BOHM: Yes, we said before that the ground was movement without division.

JK: Without division. Yes, quite.

DB: In some sense it seems inconsistent to say non-movement,
while you say the ground is movement.

JK: Yes, the ground is movement. Would you say an average, educated, sophisticated man, with all his unpleasant activities, is constantly in movement?

DB: Well, a certain kind of movement.

JK: A movement in time.

DB: Yes.

JK: A movement in becoming. But we are discussing the man who has trodden that path, if I may use that word, and come to that point. From there, what is his action? We said, for the moment, non-action, non-movement. What does that mean?

DB: It means, as you said, not taking part in this process of becoming.

JK: Of course, that is obvious. If he doesn't take part in this process, what part does he play? Is it one of complete non-action?

DB: It is not clear why you should call it non-action. We might think that it was action of another kind, which is not part of the process of becoming.

JK: It is not becoming.

DB: But it may still be action.

JK: He still has to live in the world.

DB: In one sense, whatever you do is action, but his action is not directed towards the illusory process, it is not involved in it, but would be directed towards what underlies this illusory process. It would be directed perhaps towards considering the wrong turning which is continually coming out of the ground. Right?

JK: Yes, yes. You see, various religions have described a man who has been saved, who is illuminated, who has achieved something or other. They have described very clearly, especially in Hindu religious books, how he walks, how he looks, how he talks, the whole state of his being. I think that is merely a poetic description which . . .

DB: You think it is imagination?

JK: I think a great deal of it is imagination. I have discussed this point with some people, and they say it is not like that, not imagination, that somebody who describes it knows exactly what it is.

DB: Well, how should he know? It is not clear.

JK: So what is a man of that kind? How does he live in this world? This is a very interesting question, if you go into it deeply. There is a state of non-movement. That is, the non-movement which we have gone into.

DB: You see, it is not clear exactly what you mean by "non-movement."

JK: One becomes poetic but I am trying to avoid that! Although it would be right, even poetically: It is like a single tree in a field. There is no other tree, but that tree, whatever the name of that tree is, it is there.

DB: But why do you say "non-movement"?

JK: It is non-moving.

DB: The tree stands of course.

JK: A tree is a living, moving thing. I don't mean that.

DB: The tree in a sense is moving, but in relation to the field it stands. That is the picture we get.

JK: You see, someone comes to you, because you have gone from the beginning to the end. And now you are at the end with a totally different kind of movement, which is timeless and all that. You are in that. I come to you and ask, "What is that state of mind? What is the state of your mind that has walked on that path and ended something, that has totally moved out of darkness?"

DB: If you say it is non-movement, are you implying that it is constant?

JK: It must be. . . . But what do you mean by "constant"? Continuous?

DB: No, no.

JK: Do you mean that it is . . . ?

QUESTIONER: . . . static?

JK: Oh, no!

DB: To stand firm, to stand together as a whole. That is really its literal meaning.

JK: Is that it?

DB: That is the picture you have of the tree as well. That is the picture which the tree in the field suggests.

JK: Yes, I know. That is too romantic and poetic, and it becomes rather deceptive. It is a nice image, but let's move from it. What is that mind—the quality of that mind—that has started from the beginning and pursued the becoming and gone through all that centre of darkness which has been wiped away? That mind must be entirely different. Now, what does such a mind do, or not do, in the world which is in darkness?

DB: Surely the mind does not do a thing; it does not enter into the movement of that world.

JK: Agreed.

DB: And in a sense we say that it is constant—not fixed, but it does not move.

JK: It is not static.

DB: No, it's not static. It is constant, which in a sense is also movement. There is a constancy which is not merely static, which is also, at the same time, movement.

JK: We said *that* movement was not the becoming movement.

DB: Yes, the ground movement, which is completely free.

JK: What has happened to that mind? Let's go into it a little bit. It has no anxiety and no fear. You see, the words "compassion" and "love"—it's beyond that. Right?

DB: But they may emerge out of this ground.

JK: The mind being nothing, not a thing, and therefore empty of knowledge. . . . Sorry, all this sounds so . . . unless we follow right from the beginning . . .

DB: You have to go through it, otherwise it makes no sense.

JK: Yes, no sense. So, empty of knowledge, would it always be acting in the light of insight?

DB: It would be pervaded, if not always, by the quality of insight.

JK: Yes, that is what I mean.

DB: Well "always" brings in time, you see.

JK: Remove the word.

DB: I would use "constantly."

JK: Yes, constantly; let's use the word "constant."

DB: It is a bit better, but not good enough.

JK: Yes. Let's use that word. It is acting constantly in that light, in that flash of insight. I think that is right. So what does that mean in one's daily life? How does one earn a livelihood?

DB: That, surely, would be another point. You would have to find a way to stay alive.

JK: Stay alive. So that is why I am asking this: As civilization grows, begging is not allowed.

DB: Is criminal. You have to find some way to stay alive.

JK: So what will he do? He has no profession, no special skill, no coin with which he can buy.

DB: Well, wouldn't it be possible for this mind to earn enough to get what is needed to stay alive?

JK: How?

Q: Why has he no skill to earn a livelihood?

JK: Why should he have skill? Why must one have skill to earn a livelihood? You say that, and another man says, "Why should I have skill of any kind?" I am just discussing, enquiring into this.

DB: Suppose you had to take care of yourself. You would need a certain skill. If you were by yourself in a cave, you know ...

JK: Ah, I don't want a cave!

DB: I know. But, whoever it is, he has to live somewhere; he needs some skill to find the food which he needs. You see, if everybody were to say no skill is needed, then the human race would perish.

JK: I am not sure.

DB: Well, what would happen then?

JK: That is what I am coming to. Skill implies, as we said, knowledge; from knowledge comes experience, and gradually one develops a skill. And that skill gives one an opportunity to earn a livelihood, either meagre or rich. But this man says, "There may be a different way of living and earning." We are used to that pattern, and he says, "Look, that may be totally wrong."

DB: It depends what you mean by skill. For example, suppose he has to drive a car. Surely that takes some skill?

JK: Yes.

DB: Is he going to do without that?

JK: I had better go carefully into the word "skill."

DB: Yes. I mean "skill" could have a bad meaning—like being very clever at getting money.

JK: So this man is not avaricious, he is not money-minded, he is not storing up for the future, he hasn't any insurance. But he has to live. When we use the word "skill" to mean driving a car ...

DB: . . . or being a carpenter. If all those skills were to vanish, life would be impossible.

JK: The whole thing would collapse.

DB: Yes.

JK: I am not sure. Do we mean that kind of skill must be denied?

DB: It couldn't mean that.

JK: No. That would be too silly.

DB: But then people become very skilful at getting other people to give them money, you see [*laughs*]!

JK: That might be the game. That may be it! As I am doing!

Q: I wish you were more skilled at that [*laughs*]!

JK: Sufficient unto the day thereof [*laughs*]!

Q: But is it that now we have made a division between living and skill, skill and working, living and earning a livelihood?

JK: That's it! I need to have food, I need to have clothes and shelter.

Q: But is the division necessary? As society is built now, we have a division between living and working.

JK: We have been through all that. We are talking of a man who has been through all this and has come back to the world and says, "Here I am." What is his relationship to society, and what is he to do? Has he any relationship to society?

DB: Well, not in a deep or fundamental sense, although there is a superficial relationship that he has to have.

JK: All right. A superficial contact with the world.

DB: He has to obey the laws, he has to follow the traffic signals.

JK: Quite. But I want to find out: What is he to do? Write? Talk? That means skill.

DB: Surely that kind of skill need not be harmful?

JK: I am just asking.

DB: Like the other skills; like carpentry.

JK: Yes. That kind of skill. But what is he to do? I think if we could find out the quality of a mind that has been through everything from the beginning to the end, all that we have talked of in our recent discussions, that man's mind is entirely different, yet he is in the world. How does he look upon it? You have reached and come back—these are approximate terms—and I am an ordinary man, living in this world. So what is your relationship to me? Obviously none, because I am living in a world of darkness and you are not. So your relationship can only exist when I come out of it—when darkness ends.

DB: Yes.

JK: Then there is only that; there is not a relationship. But now there is division between you and me. And I look at you with my eyes, which are accustomed to darkness and to division. But you are not. And yet you have to have some contact with me. You have to have, however superficial, however slight, a certain relationship with me. Is that relationship compassion, and not something translated by me as compassion? From my darkness I cannot judge what compassion is. Right?

DB: Yes. That follows from that.

JK: I don't know what your love is, what your compassion is, because my only love and compassion has been this. And so what do I do with you?

DB: Who are we talking about now? It is not clear to me whom we are discussing!

JK: You or X, have been through all that and come back.

DB: Then why hasn't Y done so?

JK: And Y has not. Y asks, "Who are you? You seem so different. Your way of looking at life is different." And what will Y do with X? That is the question. Not what will X do to Y. I don't know if I am making it clear.

DB: Yes, I understand. What will Y do with X?

JK: Our question until now has been what will X do with Y, but I think we were putting the wrong question. What will Y do with X? I think what would happen generally is that Y would worship, kill, or neglect him. Right?

DB: Yes.

JK: If Y worships X, then everything is very simple. He has the "goodies" of the world. But that doesn't answer my question. My question is not only what will Y do to X, but what will X do with Y? X's demand is "Look, walk out of this darkness; there is no answer in the darkness, so walk out." It doesn't matter, whatever phrase we use—walk out, dispel it, get rid of it, and so on. And Y then says, "Help me. Show me the way," and is back again in darkness. You follow? So what will Y do to X?

DB: I can't see that Y can do very much, except what you mentioned: to worship or to do something else.

JK: To kill or neglect X.

DB: But if compassion works in X . . .

JK: Yes, X is that. He won't even call it compassion.

DB: No, but we call it that. Then X will work to find a way to penetrate the darkness.

JK: Wait! So X's job is to work on darkness?

DB: To discover how to penetrate darkness.

JK: In that way he is earning a living.

DB: Well, possibly.

JK: No. I am talking seriously.

DB: It depends on whether people are willing to pay him for it.

JK: No joking. Seriously.

DB: It is possible.

JK: Probably X is the teacher. X is out of society. X is unrelated to this field of darkness and saying to the people who are caught in it, "Come out." What's wrong with that?

DB: Nothing is wrong with that.

JK: That is his means of livelihood.

DB: It's perfectly all right as long as it works. Of course, if there were a lot of people like X, there would have to be some limit.

JK: No, sir. What would happen if there were lots of people like X?

DB: That is an interesting question. I think there would be something revolutionary.

JK: That's just it.

DB: The whole framework would change.

JK: Yes. If there were lots of people like that, they would not be divided. That is the whole point, right?

DB: I think that even if ten or fifteen people were undivided, they would exert a force that has never been seen in our history.

JK: Tremendous! That's right.

DB: Because I don't think it has ever happened, that ten people have been undivided.

JK: That is X's job in life. He says that is the only thing. A group of those ten Xs will bring a totally different kind of revolution. Will society stand for that?

DB: They will have this extreme intelligence, and so they will find a way to do it, you see.

JK: Of course.

DB: Society will stand for it, because the Xs will be intelligent enough not to provoke society, and society will not react before it is too late.

JK: Quite right. You are saying something which is actually happening. Would you say then that the function of many Xs is to awaken human beings to that intelligence which will dispel the darkness? And that this is X's means of livelihood?

DB: Yes.

JK: Then there are those people who in darkness cultivate this and exploit people, but there are Xs who don't exploit. All right. That seems very simple, but I don't think it is all that simple.

DB: Right.

JK: Is that the only function of X?

DB: Well, it is really a difficult function.

JK: But I want to find out something much deeper than mere function.

DB: Yes, function is not enough.

JK: That's it. Apart from function, what is he to do? X says to Y, "Listen." And Y takes time, and gradually, perhaps, at some time he will wake up and move away. And is that all X is going to do in life?

DB: That can only be an outcome of something deeper.

JK: The deeper is all that, the ground.

DB: Yes, the ground.

JK: But is that all he is to do in this world? Just to teach people to move out of darkness?

DB: Well, that seems to be the prime task at the moment, in the sense that if this doesn't happen, the whole society will sooner or later collapse. We could ask whether he needs to be in some sense more deeply creative.

J K: What is that?

D B: Well, it is not clear.

J K: Suppose X is you, and you have an enormous field in which to operate, not merely teaching me but having this extraordinary movement which is not of time. That is, you have this abounding energy, and you have reduced all that to teach me to come out of darkness.

D B: That can only be a part of it.

J K: So what does the rest do? You follow? I don't know if I am conveying this.

D B: Well, this is what I tried to suggest by talking of some creative action, beyond this, taking place.

J K: Yes, beyond this. You may write, you may preach, you may heal, you may do this and that, but all those activities are rather trivial. But you have something else. Have I reduced you, X, to my pettiness? You can't be so reduced. My pettiness says, "You must do something. You must preach, write, heal, do something to help me to move." Right? You comply to the very smallest degree, but you have something much more than that, something immense. You understand my question?

D B: Yes. So what happens?

J K: How is that immensity operating on Y?

D B: Are you saying that there is some more direct action?

J K: Either there is more direct action, or X is doing something totally different to affect the consciousness of man.

D B: What could this be?

JK: Because X is not "satisfied" with merely preaching and talking. That immensity which he is must have an effect, must do something.

DB: Are you saying "must" in the sense of the feeling of needing to do it, or are you saying "must" in the sense of necessity?

JK: It must.

DB: It must necessarily do so. But how will it affect mankind? You see, when you say this, it would suggest to people that there is some sort of extrasensory effect that spreads.

JK: That is what I am trying to capture.

DB: Yes.

JK: That is what I am trying to convey.

DB: Not merely through the words, through the activities or gestures.

JK: Let's leave the activity alone. That is simple. It is not just that, because that immensity must . . .

DB: . . . necessarily act? There is a more direct action?

JK: No, no. All right. That immensity necessarily has other activities.

DB: Other activities at other levels?

JK: Yes, other activities. This has been translated in the Hindu teachings as various degrees of consciousness.

DB: There are different levels or degrees of acting.

JK: All that too is a very small affair. What do you say, sir?

DB: Well, since the consciousness emerges from the ground, this activity is affecting all mankind from the ground.

JK: Yes.

DB: You see, many people will find this very difficult to understand.

JK: I am not interested in many people. I want to understand— you, X, and me, Y—that that ground, that immensity, is not limited to such a petty little affair. It couldn't be.

DB: The ground includes physically the whole universe.

JK: Yes, the whole universe, and to reduce all that to . . .

DB: . . . these little activities . . .

JK: . . . is all so silly.

DB: I think that raises the question of what is the significance of mankind in the universe, or in the ground?

JK: Yes, that's it.

DB: Because even the best of these little things that we have been doing have very little significance on that scale. Right?

JK: Yes, this is just opening the chapter. I think that X is doing something—not doing, but by his very existence . . .

DB: . . . he is making something possible?

JK: Yes. When you read of Einstein, he has made something possible, which man hadn't discovered before.

DB: We can see that fairly easily because it works through the usual channels of society.

JK: Yes, I understand that. What is X bringing apart from the little things? Putting it into words makes it sound wrong. X has that immense intelligence, that energy, that something, and he must operate at a much greater level than one can possibly conceive, which must affect the consciousness of those who are living in darkness.

DB: Possibly so. The question is will this effect show in any way? You know, manifestly.

JK: Apparently not. If you hear the television or radio news and know what is happening all over the world, apparently it is not doing so.

DB: That is what is difficult, and a matter of great concern.

JK: But it must have an effect. It has to.

DB: Why do you say it has to?

JK: Because light must affect darkness.

DB: Perhaps Y might say that; living in darkness, he is not sure that there is such an effect. He might say perhaps there is, but I want to see it manifest. Not seeing anything and still being in darkness, he then asks, "What shall I do?"

JK: I understand that. So are you saying that X's only activity is just writing, teaching, and so on?

DB: No. Merely that it may well be that the activity is much greater, but it doesn't show. If only we could see it!

JK: How would it be shown? How would Y, who wants proof of it, see it?

DB: Y might say something like this: Many people have made a similar statement, and some of them have obviously been wrong. But one wants to say it could be true. You see, until now, I think the things we have said make sense, and they follow to a certain extent.

JK: Yes, I understand all that.

DB: And now you say something which goes much further. Other people have said things like that and one feels that they were on the wrong track, that they, or at least some of these people, were fooling themselves.

JK: No. X says, we are being very logical.

DB: Yes, but at this stage logic will not carry us any further.

JK: It is very reasonable! We have been through all that. So X's mind is not acting in an irrational way.

DB: You could say that, having seen the thing was reasonable, so far, Y may have some confidence that it could go further.

JK: Yes, that is what I am trying to say.

DB: Of course, there is no proof.

JK: No.

DB: So could we explore?

JK: That is what I am trying to do.

Q: What about the other activities of X? We said he has the function of teaching, but also that X has other activities.

JK: He must have. Necessarily must.

Q: But what?

JK: I don't know; we are trying to find that out.

DB: You are saying that somehow he makes possible an activity of the ground in the whole consciousness of mankind which would not have been possible without him.

JK: Yes.

Q: His contact with Y is not only verbal. Y listens but there is some other quality . . .

JK: Yes, but X says all that is a petty little affair. That is, of course, understood, but X says there is something much greater.

Q: The effect of X is perhaps far greater than can be put in words.

JK: We are trying to find out what that greater is that must necessarily be operating.

Q: Is it something that appears in the daily life of X?

JK: Yes. In his daily life X is apparently doing fairly small things—teaching, writing, bookkeeping, or whatever. But is that all? It seems so silly.

DB: Are you saying that in the daily life X does not look so different from anybody else?

JK: No, apparently not.

DB: But there is something else going on which does not show. Right?

JK: That's it. When X talks it may be different, he may say things differently but . . .

DB: . . . that is not fundamental, because there are so many people who say things differently from others.

JK: I know. But the man who has walked through all that right from the beginning! If such a man has the whole of that energy to call upon, to reduce it all to these petty little things seems ridiculous.

DB: Let me ask a question. Why does the ground require this man to operate on mankind? Why can't the ground, as it were, operate directly on mankind to clear things up?

JK: Ah, just a minute, just a minute. Are you asking why the ground demands action?

DB: Why does it require a particular man to affect mankind?

JK: Oh, that I can easily explain. It is part of existence, like the stars.

Q: Can the immensity act directly on mankind? Does it have to inform a man to enter the consciousness of mankind?

JK: We are talking about something else. I want to find out if X says, "I am not going to be reduced only to writing and talking; that is too small and petty." And the other question is why does the ground need this man? It doesn't need him.

DB: But when he is here, the ground will use him.

JK: That is all.

DB: Well, would it be possible that the ground could do something to clear this up?

JK: That is what I want to find out. That is why I am saying, in different words, that the ground doesn't need the man, but the man has touched the ground.

DB: Yes.

JK: So the ground is using him, let's say employing him. He is part of that movement. Is that all? Do you follow what I mean? Am I asking the wrong questions? Why should he do anything? Except this?

DB: Well, perhaps he does nothing.

JK: That very doing nothing may be the doing.

DB: Doing nothing makes possible the action of the ground. It may be that. In doing nothing which has any specified aim . . .

JK: That's right. No specified content which can be translated into human terms.

DB: Yes, but still he is supremely active in doing nothing.

Q: Is there an action which is beyond time, for that man?

JK: He is that . . .

Q: Then we cannot ask for a result from that man.

JK: He is not asking for results.

Q: But Y is asking for a result.

JK: No. X says, if I am concerned only to talk, that is a very small thing. But there is a vast field which must affect the whole of mankind.

DB: There is an analogy which may not be very good but we can consider it. In chemistry, a catalyst makes possible a certain action without itself taking part, merely by being what it is.

JK: Yes. Is that what is happening? Even that is a small affair.

DB: Yes.

Q: And even there Y would say it isn't happening, because the world is still in a mess. So is there a proof in the world for the activity of that man?

JK: X says he is sorry, but that is no question at all. I am not interested in proving anything. It isn't a mathematical or a technical problem to be shown and proved. X says that he has walked from the beginning of man to the very end of man, and that there is a movement which is timeless, the ground which is the universe, the cosmos, everything. And the ground doesn't need the man, but the man has come upon it. And he is still a man in the world, who says, "I write and do something or other," not to prove the ground, not to do anything. X does that just out of compassion. But there is a much greater movement which necessarily plays a part in the world.

Q: Does the greater movement play a part through X?

JK: Obviously, X says that there is something else operating which cannot possibly be put into words. He asks, "What am I to do?" There is nothing which a man like Y will understand. He will immediately translate it into some kind of illusory thing. But X says there is something else. Otherwise it is all so childish.

DB: I think the general view which people are developing now is that the universe has no meaning, that it moves any old way, things just happen, and none of them has any meaning.

JK: None of them has meaning for the man who is here, but the man who is there, speaking relatively, says it is full of meaning and not invented by thought.

All right, let's leave the vastness and all that. X says, perhaps there will be ten people with this insight and that might affect society. It will not be communism, socialism, this or that political reorganization. It might be totally different, and based on intelligence and compassion.

DB: Well, if there were ten, they might find a way to spread this much more.

JK: That's what I am trying to get at. I can't get it.

DB: What do you mean?

JK: X brings the universe, but I translate it into something trivial.

DB: Are you saying that if the whole of mankind were to see this, that would be something different?

JK: Oh, yes, of course!

DB: Would it be new ... ?

JK: It would be paradise on earth.

DB: It would be like an organism of a new kind.

JK: Of course. But, you see, I am not satisfied with this [*laughs*].

DB: Well, what is it that you ... ?

JK: I am not "satisfied" in leaving this immensity to be reduced to some few words. It seems so stupid, so incredible. You see, man Y is concerned with concepts like "Show me," "Prove it to me," "What benefit has it?," "Will it affect my future?" You follow? He is concerned with all that. And he is looking at X with eyes that are accustomed to this pettiness! So he reduces that immensity to his pettiness and puts it in a temple and has therefore lost it completely. But X says, I won't even look at that; there is something so immense, please do look at it. But Y is always translating it by wanting demonstration, proof, or reward. He is always concerned with that. [*Pause*] X brings light. That's all he can do. Isn't that enough? I think we had better stop there.

DB: To bring the light which would allow other people to be open to the immensity?

JK: You see, is it like this? We only see a small part, but that very small part extends to infinity. That means endless.

DB: Endless, yes. Small part of what?

JK: We see immensity only as a very small thing. And that immensity is the whole universe. I can't help but think that it must have some immense effect on Y, on society.

DB: Certainly the perception of this must have an effect, but it seems that it is not in the consciousness of society at the moment.

JK: I know.

DB: But you are saying that still the effect is there?

JK: Yes.

Q: Are you saying that the perception of even a small part is the infinity?

JK: Of course, of course.

Q: Is it in itself the changing factor?

JK: I think we had better stop here.

DB: Do you think it is possible that a thing like this could divert the course of mankind away from the dangerous path it is taking?

JK: Yes, that is what I think. But to divert the course of man's destruction somebody must listen. Right? Somebody—ten people—must listen!

DB: Yes.

JK: Listen to that immensity calling.

DB: So the immensity may divert the course of man. The individual cannot do it.

JK: Yes. The individual cannot do it, obviously. But X, who is supposed to be an individual, has trodden this path and says, "Listen." But man does not listen.

DB: Well, then, is it possible to discover how to make people listen?

JK: No, then we are back!

DB: What do you mean?

JK: Don't act; you have nothing to do.

DB: What does it mean not to do a thing?

JK: Thing being thought. I realize, as Y, that whatever I do—whether I sacrifice, practise, renounce—whatever I do, I am still living in that circle of darkness. So X says, "Don't act; you have nothing to do." You follow? But that is translated by Y as "That's all right. I'll wait. You do everything. I'll sit, wait, and see what happens." We must pursue this, sir, otherwise it is all so hopeless from the point of view of Y. Not to X.

Nine

Senility and the Brain Cells

1 JUNE 1980, BROCKWOOD PARK, HAMPSHIRE

JIDDU KRISHNAMURTI: I would like to talk over with you, and perhaps with Narayan too,[4] what is happening to the human brain. We have a civilization that is highly cultivated and yet at the same time barbarous, with selfishness clothed in all kinds of spiritual garb. Deep down, however, there is a frightening selfishness. Man's brain has been evolving through millennia upon millennia, yet it has come to this divisive, destructive point, which we all know. So I am wondering whether the human brain—not a particular brain but the human brain—is deteriorating. Whether it is just in a slow and steady decline. Or whether it is possible in one's lifetime to bring about in the brain a total renewal from all this, a renewal that will be pristine, original, unpolluted. I have been wondering about this, and I would like to discuss it.

I think the human brain is not a particular brain; it doesn't belong to me or to anyone else. It is the human brain which has evolved over millions of years. And in that evolution it has gathered tremendous experience, knowledge, and all the cruelties,

---

4. Mr. G. Narayan, principal of the Rishi Valley School, India.

vulgarities, and brutalities of selfishness. Is there a possibility of its sloughing off all this and becoming something else? Because apparently it is functioning in patterns. Whether it is a religious pattern, a scientific, a business, or a family pattern, it is always operating, functioning in a very small, narrow circle. Those circles are clashing against each other, and there seems to be no end to this. So what will break down this forming of patterns, so that there is no falling into other new patterns but breaking down the whole system of patterns, whether pleasant or unpleasant? After all, the brain has had many shocks, challenges, and pressures upon it, and if it is not capable of renewing or rejuvenating itself, there is very little hope. You follow?

DAVID BOHM: You see, one difficulty might present itself. If you are thinking of the brain structure, we cannot get into the structure physically.

JK: Physically we cannot. I know. We have discussed this. So what is it to do? The brain specialists can look at it, take the dead brain of a human being and examine it, but it doesn't solve the problem. Right?

DB: No.

JK: So what is a human being to do, knowing it cannot be changed from outside? The scientist, the brain specialist, and the neurologist explain various things, but their explanations, their investigations, are not going to solve this.

DB: Well, there is no evidence that they can.

JK: No evidence.

DB: Some people who do biofeedback think that they can influence the brain, connecting an instrument to the electrical potentials in the skull and being able to look at them; you can also change your heartbeat and blood pressure and other things. These people have raised the hope that something could be done.

JK: But they are not succeeding.

DB: They are not getting very far.

JK: And we can't wait for these scientists and biofeedbackers— sorry!—to solve the problem. So what shall we do?

DB: The next question is whether the brain can be aware of its own structure.

JK: Can the brain be aware of its own movement? And can the brain not only be aware of its own movement but itself have enough energy to break all patterns and move out of them?

DB: You have to ask to what extent the brain is free to break out of patterns.

JK: What do you mean?

DB: Well, you see, if you begin by saying that the brain is caught in a pattern, it may not be.

JK: Apparently it is caught.

DB: As far as we can see. It may not be free to break out. It may not have the power.

JK: That is what I have said: not enough energy, not enough power.

DB: Yes, it may not be able to take the action needed to get out.

JK: So it has become its own prisoner. Then what?

DB: Then that is the end.

JK: Is that the end?

DB: If that is true, then that is the end. If the brain cannot break out, then perhaps people would choose to try some other way to solve the problem.

NARAYAN: When we speak of the brain, in one sense it is connected to the senses and the nervous system; the feedback is there. Is there another instrument to which the brain is connected which has a different effect on the brain?

JK: What do you mean by that? Some other factor?

N: Some other factor in the human system itself. Because, obviously, through the senses the brain does get nourishment, but still that is not enough. Is there some other internal factor which gives energy to the brain?

JK: You see, I want to discuss this. The brain is constantly occupied with various problems, with holding on, attachment, and so on. It is constantly in a state of occupation. That may be the central factor. And if it is not in occupation, does it go sluggish? If it is not occupied, can it maintain the energy that is required to break down the patterns?

DB: Now, the first point is that if the brain is not occupied, somebody might think that it would just take things easy.

JK: Become lazy and all that! I don't mean that.

DB: If you mean not occupied but still active . . .

JK: Of course. I mean that.

DB: Then we have to go into what is the nature of the activity.

JK: Yes. This brain is so occupied with conflicts, struggles, attachments, fears, and pleasures. And this occupation gives to the brain its own energy. If it is not occupied, will it become lazy, drugged, and so lose its elasticity, as it were? Or will that unoccupied state give the brain the required energy to break the patterns?

DB: What makes you say this might happen? We were discussing the other day that when the brain is kept busy with intellectual activity and thought, it does not decay and shrink.

JK: As long as it is thinking, moving, living.

DB: Thinking in a rational way—then it remains strong.

JK: Yes. That is what I want to get at too. Which is as long as it is functioning, moving, thinking rationally . . .

DB: . . . it remains strong. If it starts irrational movement, then it breaks down. Also if it gets caught in a routine, it begins to die.

JK: That's it. If the brain is caught in any routine—the meditation routine or the routine of the priests.

DB: Or the daily life of the farmer . . .

JK: . . . the farmer and so on, it must gradually become dull.

DB: Not only that, but it seems to shrink physically.

JK: Yes.

DB: Perhaps some of the cells die.

JK: It shrinks physically, and the opposite to that is the eternal occupation with business—by anyone who does a routine job . . . thinking, thinking, thinking! And we believe that that also prevents shrinking.

DB: Surely experience seems to show that it does, from measurements that have been made.

JK: Yes, it does. That's it.

DB: The brain starts to shrink at a certain age. That is what they have discovered, and just as when the body is not being used, the muscles begin to lose their flexibility . . .

JK: So take lots of exercise!

DB: Well, they say exercise the body and exercise the brain.

JK: Yes. If it is caught in any pattern, any routine, any directive, it must shrink.

DB: Could we go into what makes it shrink?

JK: That is fairly simple. It is repetition.

DB: Repetition is mechanical and doesn't really use the full capacity of the brain.

JK: One has noticed that people who have spent years and years in meditation are the dullest people on earth. Also with lawyers and professors there is ample evidence of that.

N: It is suggested that rational thinking postpones senility. But rational thinking itself can sometimes become a pattern.

DB: It might. Rational thinking pursued in a narrow area might become part of the pattern too.

JK: Of course, of course.

DB: But is there some other way?

JK: We will go into that.

DB: But let's clear up things about the body first. You see, if somebody does a lot of exercise for the body, it remains strong, but it can become mechanical.

JK: Yes.

DB: And therefore it would have a bad effect.

N: What about the various traditional religious instruments—yoga, tantra, kundalini, and others?

JK: I know. Oh, they must shrink! Because of what is happening. Take yoga, for example. It used not to be vulgarized, if I may use that word. It was kept strictly to the very few, who were not concerned about kundalini and all that, but who were concerned with leading a moral, ethical, so-called spiritual life. You see, I want to get at the root of this.

DB: I think there is something related to this. It seems that before man was organized into society, he was living close to nature, and it was not possible to live in a routine.

JK: No, it was not.

DB: But it was completely insecure.

JK: So are we saying that the brain itself becomes extraordinarily alive—is not caught in a pattern—if it lives in a state of uncertainty? Without becoming neurotic?

DB: I think that is more clear when you say not becoming neurotic; otherwise uncertainty becomes a form of neurosis. But I would rather that the brain lives without having certainty, without demanding it, without demanding certain knowledge.

JK: So are we saying that knowledge also withers the brain?

DB: Yes, when it is repetitious and becomes mechanical.

JK: But knowledge itself?

DB: Well, we have to be very careful there. I think that knowledge has a tendency to become mechanical. That is, it gets fixed, but we could always be learning, you see.

JK: But learning from a centre, learning as an accumulative process!

DB: Learning with something fixed. You see, we learn something as fixed, and then you learn from there. If we were to be learning without holding anything permanently fixed ...

JK: Learning and not adding. Can we do that?

DB: Yes, I think to a certain extent we have to drop our knowledge. You see, knowledge may be valid up to a point, and then it ceases to be valid. It gets in the way. You could say that our civilization is collapsing because of too much knowledge.

JK: Of course.

DB: We don't discard what is in the way.

N: Many forms of knowledge are additive. Unless you know the previous thing, you can't do the next thing. Would you say that kind of knowledge is repetitive?

DB: No. As long as you are learning. But if you hold some principle, or the centre, fixed, and say it cannot change, then that knowledge becomes mechanical. But, for example, suppose you have to make a living. People must organize society and so on, and they need knowledge.

JK: But there we add more and more.

DB: That's right. We may also get rid of some.

JK: Of course.

DB: Some gets in the way, you see. It is continually moving there.

JK: Yes, but I am asking, apart from that, about knowledge itself.

DB: Do you mean knowledge without this content?

JK: Yes. The knowing mind.

DB: Which merely wants knowledge. Is that what you are saying? Knowledge for its own sake?

JK: Yes. I want to question the whole idea of having knowledge.

DB: But, again, it is not too clear, because we accept that we need some knowledge.

JK: Of course, at a certain level.

DB: So it is not clear what kind of knowledge it is that you are questioning.

JK: I am questioning the experience that leaves knowledge, that leaves a mark.

DB: Yes, but what kind of mark? A psychological mark?

JK: Psychological, of course.

DB: You are questioning this, rather than knowledge of technique and matter and so on. But, you see, when you use the word "knowledge" by itself, it tends to include the whole.

JK: We have said that knowledge at a certain level is essential; there you can add and take away and keep on changing. But I am questioning whether psychological knowledge is not in itself a factor of the shrinking of the brain.

DB: What do you mean by psychological knowledge? Knowledge about the mind, knowledge about myself?

JK: Yes. Knowledge about myself and living in that knowledge and accumulating that knowledge.

DB: So if you keep on accumulating knowledge about yourself or about relationships . . .

JK: Yes, about relationships. That's it. Would you say such knowledge helps the brain or makes the brain somewhat inactive, makes it shrink?

DB: Brings it into a rut.

JK: Yes.

DB: But one should see what it is about this knowledge that makes so much trouble.

JK: What is this knowledge that makes so much trouble? In relationship, that knowledge creates trouble.

DB: Yes, it gets in the way because it fixes.

JK: If I have an image about someone, that knowledge is obviously going to impede our relationship. It becomes a pattern.

DB: Yes, the knowledge about myself and about him and how we are related makes a pattern.

JK: And therefore that becomes a routine and so it loses its energy.

DB: Yes, and it occurred to me that routine in that area is more dangerous than routine in, say, the area of daily work.

JK: That's right.

DB: And if routine in ordinary work can shrink the brain, then in that area it might do some worse thing, because it has a bigger effect.

JK: Can the brain, in psychological matters, be entirely free from this kind of knowledge? Look! I am a businessman and get into the car, bus, taxi, or tube train, and I am thinking about what I am going to do, whom I am going to meet in connection with business. My mind is all the time living in that area. Then I come home. There is my wife and children, sex and all that. That also becomes a psychological knowledge from which I am acting. So there is the knowledge of my business, and also the knowledge with regard to my wife and my reactions in relationship. These two are in contradiction, unless I am unaware of them, and just carry on. If I am aware of these two, it becomes a disturbing factor.

DB: Also people find that this is a routine. They get bored with it, and they begin to ...

JK: ... divorce, and then the whole circus begins!

DB: They may hope that by becoming occupied with something else they will get out of their boredom.

JK: Yes, by going to church and so on. Any escape is an occupation. So I am asking whether this psychological knowledge is a factor of shrinkage of the brain.

DB: Well, it could be a factor.

JK: It is.

DB: If knowledge of your profession or skill can be a factor, then this psychological knowledge is stronger.

JK: Of course. Much stronger.

N: When you say psychological knowledge, you are making a distinction between psychological knowledge and, let us say, scientific knowledge or factual knowledge?

JK: Of course, we have said that.

N: But I am a little wary of the claim that scientific knowledge and other types of factual knowledge help to extend the brain, to make it bigger. That in itself doesn't lead anywhere. Though it postpones senility.

JK: Dr. Bohm makes this very clear. Rational thinking becomes merely routine; I think logically, and therefore I have learned the trick of that, but I keep on repeating it.

N: That is what happens in most forms of rational thinking.

JK: Of course.

DB: I think that there is a dependence on being continuously faced with unexpected problems.

JK: Of course.

DB: You see, lawyers may feel that their brains will last longer, because they are presented with constantly different problems, and therefore they cannot think entirely according to routine!

JK: But just a minute! They may have different clients with different problems, but they are acting from fixed knowledge.

DB: They would say not entirely; they have got to find new facts and so on.

JK: They are not functioning entirely in routine, but the basis is knowledge—precedents and book knowledge and experience with various clients.

DB: But then you would have to say that some other more subtle degeneration of the brain takes place, not merely shrinkage.

JK: That's right. That's what I want to get at.

DB: You see, when a baby is born, the brain cells have very few cross connections; these gradually increase in number, and then, as a person approaches senility, they begin to go back. So the quality of those cross connections could be wrong. If, for example, we repeated them too often, they would get too fixed.

N: Are all the brain functions confined to rational forms, or are there some functions which have a different quality?

DB: Well, it is known that a large part of the brain deals with movement of the body, with muscles, with various organs, and so on, and this part does not shrink with age, although the part that deals with rational thought, if it is not used, does shrink. Then there may be other functions that are totally unknown; that is, very little is actually known about the brain.

JK: What we are saying is that we are only using one part of the brain. There is only partial activity, partial occupation, either rational or irrational. But as long as the brain is occupied it must be in that limited area. Would you say that?

DB: Then what will happen when it is not occupied? We can say that it may tend to spend most of the time occupied in the limited set of functions which are mechanical, and that this will produce some subtle degeneration of the brain tissue, since anything like that will affect the brain tissue.

JK: Are we saying that senility is the result of a mechanical way of living? Of mechanical knowledge, so that the brain has no freedom, no space?

DB: That is the suggestion. It is not necessarily accepted by all the people who work on the brain. They have shown that the brain cells start to die around the age of thirty or forty at a steady rate, but this may be a factor. I don't think their measurements are so good that they can test effectively how the brain is used. You see, they are merely rough measurements, made statistically. But you want to propose that this death or degeneration of the brain cells comes from the wrong way of using the brain?

JK: That's right. That is what I am trying to get at.

DB: Yes, and there is a little bit of evidence from the scientists, although I think that they don't know very much about it.

JK: You see, scientists, brain specialists, are, if I may put it simply, examining things outside but not taking themselves as guinea pigs and going into that.

DB: Mostly, you see, except for those who do biofeedback, they are trying to work on themselves in a very indirect way.

JK: Yes, but I feel we haven't time for all that.

DB: It is too slow, and it isn't very deep.

JK: So let's come back to the realization that any activity which is repeated, which is directed in the narrow sense, any method, any routine, logical or illogical, does affect the brain. We have understood that very clearly. Knowledge at a certain level is essential, but psychological knowledge about oneself, one's experiences, and so on, becomes routine. The images I have about myself also obviously become routine, and all that helps to bring about a shrinkage of the brain. I have understood all that very clearly. And any kind of occupation, apart from the mechanical . . . no, not mechanical . . .

DB: . . . physical.

JK: Apart from physical occupation, the occupation with oneself brings about shrinkage of the brain. Now, how is this process to stop? And if it does stop, will there be a renewal?

DB: I think that some brain scientists would doubt that the brain cells could be renewed, and I don't know that there is any proof one way or the other.

JK: I think they can be renewed. That is what I want to get at.

DB: So we have to discuss that.

N: Are you implying that mind is different from the brain, that mind is distinct from the brain?

JK: Not quite.

DB: You have spoken of universal mind.

N: Mind in the sense that one has access to this mind, and it is not the brain. Do you consider that a possibility?

JK: I don't quite follow this. I would say that the mind is all in-clusive. When it is all inclusive—of brain, emotions, all that—when it is totally whole, not divisive in itself, there is a quality which is universal. Right?

N: One has access to it?

JK: Not "one." No, you can't reach it. You can't say, I have access to it.

N: I am only saying access. One doesn't possess it, but . . .

JK: You can't possess the sky!

N: No, my point is, is there a way of being open to it, and is there a function of the mind through which the whole of it can become accessible through education?

JK: I think there is. We may come to that presently if we can stick to this point. We are asking now if the brain can renew itself, reju-venate, become young again without any shrinkage at all. I think it can. I want to open a new chapter and discuss this. Psychologi-cally, knowledge that man has acquired is crippling it. The Freud-ians, the Jungians, the latest psychologist, the latest psychothera-pist, are all helping to make the brain shrink. Sorry! I don't mean to give offence . . .

N: Is there a way of forgetting this knowledge then?

JK: No, no. Not forgetting. I see what psychological knowledge is doing and I see the waste; I see what is taking place if I follow that line. It is obvious. So I don't follow that avenue at all. I discard analysis altogether. That is a pattern we have learnt, not only from the recent psychologists and psychotherapists but also through the tradition of a million years of analysis, of introspection, or of saying "I must" and "I must not," "This is right" and "That is wrong." You know, the whole process. I personally don't do it, and so I reject that whole method.

We are coming to a point, which is direct perception and immediate action. Our perception is generally directed by knowledge, by the past, which is knowledge perceiving, and with action arising, acting from that. This is a factor of shrinking of the brain, of senility.

Is there a perception which is not time-binding? And so action which is immediate? Am I making myself clear? That is, as long as the brain, which has evolved through time, is still living in a pattern of time, it is becoming senile. If we could break that pattern of time, the brain has broken out of its pattern, and therefore something else takes place.

N: How does the brain break out of the pattern of time?

JK: We will come to that, but first let's see if we agree.

DB: Well, you are saying that the brain is the pattern of time, and perhaps this should be clarified. I think that what you mean by analysis is some sort of process based on past knowledge, which organizes our perception, and in which we take a series of steps to try to accumulate knowledge about the whole thing. And now you say that this is a pattern of time, and we have to break out of it.

JK: If we agree that this is so, the brain is functioning in a pattern of time.

DB: Then we have to ask, what other pattern is possible?

JK: But wait . . .

DB: What other movement is possible?

JK: No. First let's understand this, not merely verbally, but let's actually see that it is happening. That our action, our way of living, our whole thinking, is bound by time, or comes with the knowledge of time.

DB: Certainly our thinking about ourselves, any attempt to analyse ourselves, to think about ourselves, involves this process.

JK: This process, which is of time. Right?

N: That is a difficulty. When you say knowledge and experience, they are a certain cohesive energy or force that binds you.

JK: Which means what? Time-binding!

N: Time-binding and . . .

JK: . . . and therefore the pattern of centuries, of millennia, is being repeated.

N: Yes. But I am saying that this has a certain cohesive force.

JK: Of course, of course. All illusions have an extraordinary vitality.

N: Very few break through.

JK: Look at all the churches and what immense vitality they have.

N: No, apart from these churches, in one's personal life it has a certain cohesive force that keeps one back. One can't break away from it.

JK: What do you mean it keeps you back?

N: It has a magnetic attraction; it sort of pulls you back. You can't free yourself of it unless you have some instrument with which you can act.

JK: We are going to find out if there is a different approach to the problem.

DB: When you say a different instrument, that is not clear. The whole notion of an instrument involves time, because if you use any instrument, it is a process which you plan.

JK: Time. That's just it.

N: I use the word "instrument" in the sense of effective.

JK: It has not been effective. On the contrary, it is destructive. So do I see the very truth of its destructiveness? Not just the theory, the idea, but the actuality of it. If I do, then what takes place? The brain has evolved through time and has been functioning, living, acting, believing in that time process. But when one realizes that all this helps to make the brain senile, when one sees that as true, then what is the next step?

N: Are you implying that the very seeing that it is destructive is a releasing factor?

JK: Yes.

N: And there is no need for an extra instrument?

JK: No. Don't use the word "instrument." There is no other factor. We are concerned to end this shrinkage and senility, and asking whether the brain itself, the cells, the whole thing, can move out of time. I am not talking about immortality and all that kind of stuff! Can the brain move out of time altogether? Otherwise deterioration, shrinkage, and senility are inevitable, and even when senility may not show, the brain cells are becoming weaker and so on.

N: If the brain cells are material and physical, somehow or other they have to shrink through time; indeed it can't be helped. The brain cell, which is tissue, cannot in physical terms be immortal.

DB: Perhaps the rate of shrinkage would be greatly slowed down. If a person lives a certain number of years, and his brain begins to shrink long before he dies, then he becomes senile. Now, if the deterioration would slow, then . . .

JK: . . . not only slow down, sir.

DB: Well, regenerate.

JK: Be in a state of non-occupation.

DB: I think Narayan is saying that it is impossible for any material system to last forever.

JK: I am not talking about lasting forever—though I am not sure if it can't last forever! No, this is very serious. I am not pulling anybody's leg.

DB: If all the cells were to regenerate in the body and in the brain, then the whole thing could go on indefinitely.

JK: Look, we are now destroying the body, through drink, smoking, overindulgence in sex, and all kinds of things. We are living most unhealthily. Right? If the body were in excellent health, maintained right through—which means no heightened emotions, no strain on, no sense of deterioration of the body, the heart functioning normally—then why not!

DB: Well . . .

JK: Which means what? No travelling and all the rest of it . . .

DB: No excitement.

JK: If the body remains in one quiet place, I am sure it can last a great many more years than it does now.

DB: Yes, I think that is true. There have been many cases of people living for a hundred and fifty years in quiet places. I think that is all you are talking about. You are not really suggesting something lasting forever?

JK: The body can be kept healthy, and since the body affects the mind, nerves, senses, and all that, they also can be kept healthy.

DB: And if the brain is kept in the right action . . .

JK: Yes, without any strain.

DB: You see, the brain has a tremendous affect on organizing the body. The pituitary gland controls the entire system of the body glands; also all the organs of the body are controlled by the brain. When the brain deteriorates, the body starts to deteriorate.

JK: Of course.

DB: They work together.

JK: They go together. So can this brain—which is not "my" brain—which has evolved through millions of years, which has had all kinds of destructive or pleasant experiences . . .

DB: You mean it is a typical brain, not a particular brain, peculiar to some individual? When you say "not mine," you mean any brain belonging to mankind, right?

JK: Any brain.

DB: They are all basically similar.

JK: Similar: That is what I said. Can that brain be free of all this? Of time? I think it can.

DB: Perhaps we could discuss what it means to be free of time. You see, at first the suggestion that the brain be free of time might sound crazy, but, obviously, we all know that you don't mean that the clock stops.

JK: Science fiction and all that!

DB: The point is what does it really mean to be psychologically free of time?

JK: That there is no tomorrow.

DB: But we know there is tomorrow.

JK: But psychologically...

DB: Can you describe better what you mean when you say "no tomorrow"?

JK: What does it mean to be living in time? Let's take the other side first, because then we come to the other. What does it mean to live in time? Hope; thinking and living in the past, and acting from the knowledge of the past; images, illusions, prejudices— they are all an outcome of the past. All that is time, and that is producing chaos in the world.

DB: Well, suppose we say that if we are not living psychologically in time, we may still order our actions by the watch. The thing that is puzzling is if somebody says, "I am not living in time, but I must keep an appointment." You see?

JK: Of course. You can't sit here forever.

DB: So you say, I am looking at the watch, but I am not psychologically extending how I am going to feel in the next hour, when I have fulfilment of desire or whatever.

JK: I am just saying that the way we are living now is in the field of time. And there we have brought all kinds of problems and suffering. Is that right?

DB: Yes, but it should be made clear why this necessarily produces suffering. You are saying that if you live in the field of time, suffering is inevitable.

JK: Inevitable.

DB: Why?

JK: It is simple. Time has built the ego, the "me," the image of me sustained by society, by parents, by education, which has built it through millions of years. All that is the result of time. And from there I act.

N: Yes.

DB: Towards the future psychologically. That is, towards some future state of being.

JK: Yes. Which means that the centre is always becoming.

DB: Trying to become better.

JK: Better, nobler, or anything else. So all that, the constant endeavour to become something psychologically is a factor of time.

DB: Are you saying that the endeavour to become produces suffering?

JK: Obviously. It is simple. All that is divisive. It divides me from others, and so you are different from me. And when I depend on somebody, and that somebody is gone, I feel lonely and miserable. All that goes on. So we are saying that any factor of division, which is the very nature of the self, must inevitably cause suffering.

DB: Are you saying that through time the self is set up, and then the self introduces division and conflict and so on? But that if there were no psychological time, then perhaps this entire structure would collapse, and something entirely different would happen?

JK: That's it. That is what I am saying. And therefore the brain itself has broken down.

DB: Well, that is the next step—to say that the brain has broken out of that rut and perhaps could then regenerate. It doesn't follow logically, but still it could be so.

JK: I think it does follow logically.

DB: Well, it follows logically that it would stop degenerating.

JK: Yes.

DB: And are you adding further that it would start to regenerate?

JK: You look sceptical.

N: Yes, because the whole human predicament is bound to time.

JK: We know that.

N: Society, individuals, the whole structure.

JK: I know, I know.

N: It is so forceful that anything feeble doesn't work here.

JK: What do you mean "feeble"?

N: The force of this is so great that what has to break through must have greater energy.

JK: Yes.

N: And no individual seems to be able to generate sufficient energy to be able to break through.

JK: But you have got hold of the wrong end of the stick, if I may point out. When you use the word "individual," you have moved away from the fact that our brain is universal.

N: Yes, I admit that.

JK: There is no individuality.

N: That brain is conditioned this way.

JK: Yes, we have been through all that. It is conditioned this way through time. Time is conditioning, right? It is not that time has created the conditioning; time itself is the factor of conditioning.

So can that time element not exist? We are talking about psychological time, not the ordinary physical time. I say it can. We have said that the ending of suffering comes about when the self, which is built up through time, is no longer there. A man who is actually going through agony might reject this, is bound to. But when he comes out of the shock of it, if somebody points out to him what is happening, and if he is willing to listen, to see the rationality, the sanity of it, and not to build a wall against it, he is out of that field. The brain is out of that time-binding quality.

N: Temporarily.

JK: Ah! There again when you use the word "temporary," it means time.

N: No, I mean that the man slips back into time.

JK: No, he can't. He can't go back if he sees that something is dangerous, like a cobra or any other danger. He cannot go back to it.

N: That analogy is a bit difficult, because the structure itself is that danger. One inadvertently slips into it.

JK: Look, Narayan, when you see a dangerous animal, there is immediate action. It may be the result of past knowledge and experience, but there is immediate action for self-protection. But psychologically we are aware of the dangers. If we become as aware of these dangers as we are aware of physical dangers, there is an action which is not time-binding.

DB: Yes, I think you could say that as long as you could perceive this danger, you know you would respond immediately. But, you see, if you were to use that analogy of the animal, it might be an animal that you realize is dangerous, but it might take another form that you don't see as dangerous!

JK: Yes.

DB: Therefore there would be a danger of slipping back if you didn't see that this time illusion might come in some other form.

JK: Of course.

DB: But I think the major point you are making is that the brain does not belong to any individual.

JK: Yes, absolutely.

DB: And therefore it is no use saying that the individual slips back.

JK: No.

DB: Because that already denies what you are saying. The danger is rather that the brain might slip back.

JK: The brain itself might slip back, because it has not seen the danger.

DB: It hasn't seen the other forms of the illusions.

JK: The Holy Ghost taking different shapes! Time is the real root of this.

DB: Time and separation as individuality are basically the same structure.

JK: Of course.

DB: Although it is not obvious in the beginning.

JK: I wonder if we see that.

DB: It might be worth discussing this. Why is psychological time the same illusion, the same structure as individuality? Individuality is the sense of being a person who is located here somewhere.

JK: Located and divided.

DB: Divided from the others. He extends out to some periphery; his domain extends out to some periphery, and also he has an identity which extends over time. He wouldn't regard himself as an individual if he said, "Today I am one person, tomorrow I am another." So it seems that we mean by "individual" somebody who is in time.

JK: I think that this idea of individuality is such a fallacy.

DB: Yes, but many people may find it very hard to be convinced that it is a fallacy. There is a common feeling that, as an individual, I have existed at least from my birth, if not before, and go on to death and perhaps later. The whole idea of being an individual is to be in time. Right?

JK: Obviously.

DB: To be in psychological time, not just the time of the clock.

JK: Yes, we are saying that. So can that illusion that time has created individuality be broken? Can this brain understand that?

DB: I think that, as Narayan said, there is a great momentum in any brain, which keeps rolling, moving along.

JK: Can that momentum stop?

N: The difficulty comes here. The genetic coding is intrinsic to a person. You seem to function more or less unconsciously, driven by this past momentum. And suddenly you see, like a flash, something true. But the difficulty is that it may operate only for a day— and then you are again caught in the old momentum.

JK: I know that. But I say the brain will not be caught. Once the mind or the brain is aware of this fact, it cannot go back. How can it?

N: There must be another way of preventing it from going back.

JK: Not preventing; that also means time. You are still thinking in terms of prevention.

N: Prevention in the sense of a human factor.

JK: The human being is irrational. Right? And as long as he is functioning irrationally, he says of any rational factor "I refuse to see it."

N: You are suggesting that the very seeing prevents you from slipping back. This is a human condition.

DB: I wonder if we should go further into this question about prevention. It may be important.

N: There are two aspects. You see the fallacy of something, and the very seeing prevents you from slipping back, because you see the danger of it.

DB: In another sense you say you have no temptation to slip back, therefore you don't have to be prevented. If you really see it, there is no need for conscious prevention.

N: Then you are not tempted to go back.

JK: I can't go back. If, for example, I see the fallacy of all the religious nonsense, it is finished!

DB: The only question which I raise is that you may not see this so completely in another form.

N: It may come in different shapes . . .

DB: And then you are tempted once again.

JK: The mind is aware; it is not caught. But you are saying that it is.

N: Yes, in other shapes and forms.

JK: Wait, sir. We have said that perception is out of time, is seeing immediately the whole nature of time. Which, to use a good old word, is to have an insight into the nature of time. If there is that insight, the very brain cells, which are part of time, break down. The brain cells bring about a change in themselves. You may disagree, you may say, "Prove it." I say this is not a matter of proof; it is a matter of action. Do it, find out, test it.

N: You were also saying the other day that when the consciousness is empty of its content . . .

JK: . . . the content being time . . .

N: . . . that leads to the transformation of the brain cells.

JK: Yes.

N: When you say consciousness is empty of the content there . . .

JK: . . . there is no consciousness as we know it.

N: Yes. And you are using the word "insight." What is the connection between the two?

DB: Between what?

N: Consciousness and insight. You have suggested that when consciousness is empty of its content . . .

JK: Be careful. Consciousness is put together by its content. The content is the result of time.

DB: The content also *is* time.

JK: Of course.

DB: It is about time as well, and it is actually put together by time; also it is about time.

JK: Now, if you have an insight into that, the whole pattern is gone, broken. The insight is not of time, not of memory, is not of knowledge.

N: Who has this insight?

JK: Not "who." Simply, there is an insight.

N: There is an insight and then the consciousness is empty of its content...

JK: No, sir. No.

N: You are implying that the very emptying of the content is insight?

JK: No, we are saying time is a factor which has made up the content. It has built it up, and it also thinks about it. All that bundle is the result of time. Insight into this whole movement, which is not "my" insight, brings about transformation in the brain. Because that insight is not time-binding.

DB: Are you saying that this psychological content is a certain structure, physically, in the brain? That in order for this psychological content to exist, the brain over many years has made many connections of the cells, which constitute this content?

JK: Quite, quite.

DB: And then there is a flash of insight, which sees all this, and that it is not necessary. Therefore all this begins to dissipate. And when it has dissipated, there is no content. Then whatever the brain does is something different.

JK: Let us go further. Then there is total emptiness.

DB: Well, emptiness of that content. But when you say total emptiness, you mean emptiness of all this inward content?

JK: That's right. And that emptiness has tremendous energy. It is energy.

DB: So could you say that the brain, having had all these connections tangled, has locked up a lot of energy?

JK: That's right. Wastage of energy.

DB: And when they begin to dissipate, that energy is there.

JK: Yes.

DB: Would you say that it is as much physical energy as any other kind?

JK: Of course. Now, we can go on in more detail, but is this principle, the root of it, an idea or a fact? I hear all this physically with the ear, but I may make it into an idea. If I hear it, not only with the ear but in my being, in the very structure of myself, what happens then? If that kind of hearing doesn't take place, all this becomes merely an idea, and I spin along for the rest of my life playing with ideas.

Now, we, more or less, are "a captive audience" here. But if there was a scientist here, biofeedback or another brain specialist, would he accept all this? Would he even listen to it?

DB: A few scientists would, but obviously the majority would not.

JK: No. So how do we touch the human brain?

DB: You see, all this will sound rather abstract to most scientists. They will say it could be so, it is a nice theory, but we have no proof of it.

JK: Of course.

DB: They would say it doesn't excite them very much because they don't see any proof. They would say if you have some more evidence, we will come back later and become very interested. And you can't give any proof because whatever is happening, nobody can see with their eyes.

JK: I understand. But I am asking: What shall we do? The human brain—not "my" brain or "you" brain—has evolved through a million years. One "biological freak" can move out of it, but how do you get at the human mind generally to make it see all this?

DB: I think you have to communicate the necessity, the inevitability of what you are saying. If a person sees something and you explain it to him, and he sees it happening before his eyes, he says, "That's so."

JK: But it requires somebody to listen, somebody who says, "I want to capture it, I want to understand this, I want to find out." You follow what I am saying? Apparently that is one of the most difficult things in life.

DB: Well, it is the function of this occupied brain—that it is occupied with itself and it doesn't listen.

N: In fact one of the things is that this occupation starts very early. When you are young, it is very powerful, and it continues all through your life. How can we, through education, make this clear?

JK: The moment you see the importance of not being occupied—see that as a tremendous truth—you will find ways and methods to help educationally, creatively. No one can be told, copy, and imitate, for then he is lost.

DB: Then the question is how is it possible to communicate to the brain which rejects, which doesn't listen? Is there a way?

JK: Not if I refuse to listen. You see, I think meditation is a great factor in all this. I feel we have been meditating, although ordinarily people wouldn't accept this as meditation.

DB: They have used the word so often . . .

JK: . . . that its meaning is really lost. But true meditation is this: the emptying of consciousness. You follow?

DB: Yes, but let's be clear. Earlier you said it would happen through insight. Now are you saying that meditation is conducive to insight?

JK: Meditation is insight.

DB: It is insight already. Then is it some sort of work you do? Insight is usually thought of as the flash, but meditation is more constant.

JK: We must be careful. What do we mean by meditation? We can reject the systems, methods, acknowledged authorities, Zen, Tibetan, Hindu, Buddhist, because this is obviously mere tradition, repetition, and time-binding nonsense.

N: Do you think some of them could have been original, could have had original insight, in the past?

JK: If they had, they wouldn't belong to Christianity, Hinduism, Buddhism. They wouldn't be anything. I mean, who knows? Now meditation is this penetration, is this sense of moving without any past.

DB: The only point to clear up is that when you use the word "meditation," you mean something more than insight, you see.

JK: Much more. Insight has freed the brain from the past, from time. That is an enormous statement . . .

DB: Do you mean that you have to have insight if you are going to meditate?

JK: Yes, that's right. To meditate without any sense of becoming.

DB: You cannot meditate without insight. You can't regard it as a procedure by which you will come to insight.

JK: No, that immediately implies time. A procedure, a system, a method, in order to have insight is nonsensical. Insight into greed or fear frees the mind from them. Then meditation has quite a different quality. It has nothing to do with all the gurus' meditations. So could we say that to have insight there must be silence?

DB: Well, that is the same; we seem to be going in a circle.

JK: For the moment.

DB: My mind has silence.

JK: So the silence of insight has cleansed, purged all that.

DB: All that structure of the occupation.

JK: Yes. So meditation—what is it? There is no movement as we know it; no movement of time.

DB: Is there movement of some other kind?

JK: I don't see how we can measure that by words, that sense of a limitless state.

DB: But you were saying earlier that nevertheless it is necessary to find some language, even though it is unsayable!

JK: Yes. We will find that language.

Ten

Cosmic Order

7 JUNE 1980, BROCKWOOD PARK, HAMPSHIRE

JIDDU KRISHNAMURTI: We left off the other day by saying that when the mind is totally empty of all the things that thought has put there, then real meditation begins. But I would like to go more deeply into that matter, to go back a bit and find out if the mind, the brain, can ever be free from all illusion, any forms of deception. Also whether it can have its own order—an order not introduced by thought, effort, or any endeavour to put things in their proper place. And also, however much damaged the brain is by shock and all kinds of stimulation, whether it can heal itself completely.

So first let's begin by asking if there is an order which is not made by man or by thought—which is not the result of calculated order out of disturbance, and therefore still part of the old conditioning.

DAVID BOHM: Are you referring to the mind? I mean you can say the order of nature exists on its own.

JK: The order of nature is order.

DB: Yes, it is not made by man.

JK: But I am not talking of such. I am not sure that it is that kind of order. Is there cosmic order?

DB: Well, that is still the same thing, in a sense, because the word "cosmos" means order, but the whole order, which includes the order of the universe and the order of the mind.

JK: Yes. What I am trying to find out is whether there is order which man can never possibly conceive? Because any concept is still within the pattern of thought.

DB: Well, how are we going to discuss it?

JK: I don't know. I think we can. What is order?

NARAYAN: There is mathematical order, the highest kind of order known to any discipline.

JK: Would the mathematicians agree that mathematics is complete order?

N: Yes, mathematics itself is order.

DB: I think it depends on the mathematician. But there is a well-known mathematician called von Neumann who defined mathematics as the relationship of relationships. Really he meant, by relationship, order. It is order working within the field of order itself, rather than working on some object.

JK: Yes, that is what I am trying to get at.

DB: So the most creative mathematicians are having a perception of this, which may be called pure order; but of course it is limited, because it has to be expressed mathematically, in terms of formulae or equations.

JK: Of course. Is order part of disorder, as we know it?

DB: What we mean by disorder is another question. It is not possible to give a coherent definition of disorder, because it violates order. Anything that actually happens has an order, but you can call a certain thing disorder if you like.

JK: Are you saying that anything that happens is order?

DB: Has some order. If the body is not functioning rightly, even if cancer is growing, there is a certain order in the cancer cell; it is just growing according to a different pattern, which tends to break down the body. Nevertheless the whole thing has a certain kind of order.

JK: Yes, yes.

DB: It has not violated the laws of nature, although relative to some context you could say it is disorder; because, if we are talking of the health of the body, then the cancer is called disorder. But in itself...

JK: Cancer has its own order.

DB: Yes, but it is not compatible with the order of the growth of the body.

JK: Quite. So what do we mean by order? Is there such a thing as order?

DB: Order is a perception; we can't get hold of order.

N: I think that generally when we refer to order, it is in relation to a framework or a certain field. Order always has that connotation. But when you say the order of order, as in the study of mathematics, we are going away from this limited approach to it.

D B: You see, most mathematics start with the order of the numbers, like one, two, three, four, and build on that, in a hierarchy. But you can see what is meant by the order of the numbers. There is,. for example, a series of relationships which are constant. In the order of the numbers, you have the simplest example of order.

N: And a new order was created with the discovery of zero! Are mathematical order and the order in nature part of a bigger field? Or are these localized forms?

J K: You see the brain, the mind, is so contradictory, so bruised, that it can't find order.

D B: Yes, but what kind of order does it want?

J K: It wants an order in which it will be safe, in which it won't be bruised, be shocked, or feel physical and psychological pain.

D B: The whole point of order and of mathematics is not to have contradiction.

J K: But the brain is in contradiction.

D B: And something has gone wrong.

J K: Yes, we have said that the brain took a wrong turn.

D B: You see, if the body is growing wrongly, we have a cancer cell, which means two contradictory orders—one being the growth of the cancer and the other the order of the body.

J K: Yes. But can the mind, the brain, be totally free of all organized order?

D B: You mean, by organized order, a fixed or imposed pattern?

JK: Yes. Imposed or self-imposed. We are trying to investigate whether the brain can ever be free from all the impositions, pressures, wounds, bruises, and trivialities of existence which are pushing it in different directions. If it cannot, meditation has no meaning.

DB: You could go further and say that probably life has no meaning if you cannot free it of all that.

JK: No, I wouldn't say that life has no meaning.

DB: If the pattern goes on indefinitely . . .

JK: If it goes on as it has done, indefinitely, for millennia, life has no meaning. But to find out if it has a meaning, and I think it does, must the brain be totally free of all this?

DB: Well, what is the source of what we call disorder? It is almost like a cancer going on inside the brain, moving in a way which is not compatible with the health of the brain.

JK: Yes.

DB: It grows as time goes on, it increases from one generation to another.

JK: Each generation repeats the same pattern.

DB: It tends to accumulate through tradition with every generation.

JK: How is this set, accumulated pattern to end, to be broken through?

DB: Could we ask another question? Why does the brain provide the soil for this stuff to grow on?

JK: It may be merely tradition or habit.

DB: But why does the brain stay in that?

JK: It feels safe. It is afraid of something new taking place, because in the old tradition it finds refuge.

DB: Then we have to question why the brain deceives itself. This pattern involves the fact that the brain deceives itself about disorder. It doesn't seem able to see it clearly.

N: In my mind there is intelligence behind order which makes use of it. I have a certain purpose for which I create an order, and when the purpose is over I set aside that order or pattern. So order has an intelligence which works it out. That is the usual connotation. But you are referring to something else.

JK: I am asking whether this pattern of generations can be broken, and why the brain has accepted that pattern in spite of all its conflicts and misery.

N: I am saying the same thing in a different way. When an order has served its purpose, can it then be put aside?

JK: Apparently it can't. We are speaking psychologically. It can't. The brain goes on repeating fears, sorrow, miseries. All that is part of its daily meal! Is it so heavily conditioned that it cannot see its way out of it? Or because, by constant repetition, the brain has become dull?

N: The momentum of repetition is there.

JK: Yes. That momentum makes the mind sluggish, mechanical. And in that mechanical sluggishness it takes refuge and says, "It's all right. I can go on." That's what most human beings do.

DB: That is part of the disorder. To think in that way is a manifestation of disorder.

JK: Of course.

N: Do you connect order with intelligence? Or is order something that exists on its own?

DB: Intelligence involves order; it requires the perception of order in an orderly way, without contradiction. But I think that, in the terms of this discussion, we ourselves don't create this; we don't impose this order, but rather it is natural.

JK: Yes. So let's come back. I am the ordinary man. I see that I am caught. My whole way of living and thinking, my attitudes and beliefs, are out of this enormous length of time. Time is my whole existence. In the past, which cannot be changed, I take refuge. Right?

DB: Well, I think that if we were to talk to the so-called ordinary man, we would find he doesn't really understand that time is something that happens to him.

JK: I am saying an ordinary man can see, after talking over with another, that his whole existence is based on time. And the mind takes refuge in time—in the past.

DB: What does that mean exactly? How does it take refuge?

JK: Because the past cannot be changed.

DB: Yes, but people also think of the future. It is common to think that the future can change. The communists have said, "Give up the past, we are going to change the future."

JK: But we can't give up the past, even if we think we can.

DB: Then if even those who try not to take refuge in the past can't give it up, it seems that whatever we do, we are stuck.

JK: So the next step is why does the brain accept this way of living? Why doesn't it break it down? Is it through laziness or that in breaking it down it has no hope?

DB: That is still the same question, of going from past to future.

JK: Of course. So what is the brain to do? This is applicable to most people, isn't it?

DB: We haven't understood why, when people see that their behaviour is disorderly or irrational, they try to give up the past but find they cannot. Why can't they?

JK: Wait, sir. If I give up the past, I have no existence. If I give up all my remembrances, I have nothing; I am nothing.

DB: I think some people like the Marxists would look at it a little differently. Marx said that it is necessary to transform the conditions of human existence and that this will remove the past.

JK: But it has not done so. It cannot be done.

DB: That is because when man tries to transform it, he still works from the past.

JK: Yes, that's what I am saying.

DB: If you say, don't depend on the past at all, then, as you have asked, what are we going to do?

JK: I am nothing: Is that the reason why we cannot possibly give up the past? Because my existence, my way of thinking, my life, *everything*, is from the past. And if you say, wipe that out, what have I left?

DB: I think you could say that obviously we have to keep certain things from the past, like useful knowledge.

JK: Yes, we have been through all that.

DB: But you could ask, suppose we keep that useful part of the past and wipe out all aspects of the past which are contradictory?

JK: Which are all psychologically contradictory. Then what is left? Just going to the office? There is nothing. Is that the reason why we cannot give it up?

DB: There is still a contradiction in that, because if you say, "What is left?" you are still asking for the past.

JK: Of course.

DB: Are you simply saying that when people talk of giving up the past, they are just not doing it, but merely turning this into another question which avoids the issue?

JK: Because my whole being is the past; it has changed or been modified, but its roots are in the past.

DB: Now, if you said, "All right, give all that up and in the future you will have something quite different, and better," would people then be attracted to this?

JK: But "better" is still from the past.

DB: But people want to be assured of at least something.

JK: That is just it. There is nothing. The ordinary human being wants something to which he can cling.

DB: He may feel not that he is clinging to the past but reaching for something.

JK: If I reach for something, it is still the past.

DB: Yes, it has its roots in the past, but that is not often obvious, because people say it is a big, new revolutionary situation.

JK: As long as I have my roots in the past there cannot be order.

DB: Because the past is pervaded with disorder.

JK: Yes. And is my mind, my brain, willing to see that there is absolutely nothing if I give up the past?

DB: And nothing to reach for.

JK: Nothing. There is no movement. So people dangle a carrot in front of me and, foolishly, I follow it. But if I have no carrots, no rewards or punishments, how is this past to be dissolved? Because otherwise I am still living in the field of time that is man-made. So what shall I do? Am I willing to face absolute emptiness?

DB: What will you tell somebody who is not willing, or feels unable, to face this?

JK: I am not bothered. If somebody else says, "I can't do all this nonsense," I say, "Well, carry on."

But I am willing to let my past go completely. Which means there is no effort or reward or punishment, no carrot, nothing. And the brain is willing to face this extraordinary and totally new state to it of existing in a state of nothingness. That is appallingly frightening.

DB: Even these words will have their meaning rooted in the past, and that's where fear comes in.

JK: Of course. We have understood that; the word is not the thing. My brain says it is willing to do that, to face this absolute nothingness and emptiness, because it has seen for itself that all the places where it has taken refuge are illusions . . .

DB: I think this leaves out something that you brought up earlier—the question of the damage or scars to the brain.

JK: That is just it.

DB: The brain that isn't damaged could possibly let go the past fairly readily.

JK: Look, can I discover what has caused damage to the brain? Surely one of the factors is strong, sustained emotions, like hatred.

DB: Probably a flash of emotion doesn't do so much damage, but people sustain it.

JK: Of course. Hatred, anger, and violence not only shock but wound the brain. Right?

DB: And getting excessively excited.

JK: Of course. And drugs and so on. The natural response doesn't damage the brain. Now the brain is damaged; suppose it has been damaged through anger.

DB: You could even say that nerves probably get connected up in the wrong way and that the connections are too fixed. I think there is evidence that these things will actually change the structure.

JK: Yes, and can we have an insight into the whole nature of disturbance, so that the insight changes the cells of the brain which have been wounded?

DB: Well, possibly it would start them healing.

JK: All right, start them healing. That healing must be immediate.

DB: It may take time in the sense that if wrong connections have been made, it is going to take time to redistribute the material. The beginning of it, it seems to me, is immediate.

JK: All right. Can I do that? I have listened to X, I have carefully read, I have thought about all this, and I see that anger, violence, hatred—any excessive emotion—bruises the brain. And insight into this whole business brings about a mutation in the cells. It is so. Also the nerves, and all their adjustments, will be as rapid as possible.

DB: Something like this happens with cancer cells. Sometimes the cancer suddenly stops growing, and it goes the other way, for some reason that is unknown. But a change must have taken place in those cells.

JK: Could it be that the brain cells change fundamentally, and the cancer process stops?

DB: Yes. Fundamentally it stops and begins to dismantle.

JK: Dismantle. Yes, that is it.

N: You are saying that insight sets in motion the right kind of connections and stops the wrong connections?

DB: And even dismantles the wrong connections.

N: So a beginning is made, and it is made now.

DB: At one moment.

JK: That is the insight.

N: But there is no time involved, because the right movement has started now. There is another thing which I want to ask about the past: For most people, the past means pleasure.

JK: Not only pleasure but the remembrance of everything.

N: One starts disliking pleasure only when it becomes stale or leads to difficulties. One wants pleasure all the time.

JK: Of course.

N: It is sometimes difficult to distinguish between pleasure and the staleness or the difficulties that it brings.

JK: Pleasure is always the past; there is no pleasure at the moment it is happening. That comes in later, when it is remembered. So the remembrance is the past. But I am willing to face nothingness, which means to wipe out all that!

N: But I am saying that the human being, even though he understands what you are saying, is held back in this field.

JK: Because he is not willing to face this emptiness. Pleasure is not compassion, pleasure is not love, pleasure has no place in compassion. But perhaps if there is this mutation, compassion is stronger than pleasure.

DB: Even the perception of order may be stronger than pleasure. If people are really concerned with something, the pleasure plays no role at that moment.

N: But what happens to a man in whom pleasure is dominant?

JK: We have already discussed this. As long as he is unwilling to face this extraordinary emptiness, he will keep on with the old pattern.

DB: You see, we have to say that this man had a damaged brain too. It is brain damage which causes this emphasis on sustained pleasure, as well as the fear and the anger and the hate . . .

JK: But the damaged brain is healed when there is insight.

DB: Yes, but I think many people who would understand that hate and anger are products of the damaged brain would find it hard to see that pleasure is also the product of the damaged brain.

JK: Oh, yes, but of course it is.

DB: Can we say there is a true enjoyment, which is not the product of the damaged brain, which is confused with pleasure . . . ?

N: If pleasure gives rise to anger, anger is part of the damaged brain.

JK: And also the demand for pleasure. So do you have an insight into how very destructive the past is to the brain? Can the brain itself see, have an insight into this, and move out of it?

N: You are saying that the beginning of order comes from insight?

JK: Obviously. Let's work from there.

N: May I put it in a different way? Is it possible to gather a certain amount of order in a patterned sense, not artificially, so that it gives rise to a certain amount of insight?

JK: Ah! You cannot find truth through the false.

N: I am asking it purposefully because many people seem to lack the energy that is required for insight.

JK: You are tremendously keen to earn a livelihood, to earn money, to do anything in which you are really interested. If you are interested vitally in this transformation, you have the energy.

May we go on? I, as a human being, have seen that this insight has wiped away the past, and the brain is willing to live in nothingness. Right? We have come to this point several times from different directions. Now, let's go on. Now, there isn't a thing put there by thought. There is no movement of thought, except for technical thought, knowledge, which has its own place. But we are talking of the psychological state of mind where there is no movement of thought. There is absolutely nothing.

DB: You mean no feeling either? You see, the movement of thought and feeling is together.

JK: Wait a minute. What do you mean here by feeling?

DB: Well, usually people might say, all right, there is no thought, but they have various feelings.

JK: Of course we have feelings. The moment you stick a pin in me . . .

DB: Those are sensations. And also there are the inner feelings.

JK: Inner feelings of what?

DB: It is hard to describe them. Those that can easily be described are obviously the wrong kind, such as anger and fear.

JK: Is compassion a feeling?

DB: Probably not.

JK: No, it is not a feeling.

DB: Though people may say they *feel* compassionate! Even the very word suggests it is a form of feeling. Compassion has in it the word "passion," which is feeling. This is a difficult question. We could perhaps question what we usually recognize as feelings.

JK: Let's go into that a little bit. What do we mean by feelings? Sensations?

DB: Well, people don't usually mean that. You see, sensations are connected with the body.

JK: So you are talking of feelings which are not of the body?

DB: Yes, or which—in the old days—would have been described as of the soul.

JK: The soul, of course. That is an easy escape, but it means nothing.

DB: No.

JK: What are the inner feelings? Pleasure?

DB: Well, insofar as you can label it that way, it is clear that it is not valid.

JK: So what is valid? The nonverbal state?

DB: It may be a nonverbal state . . . something analogous to a feeling which isn't fixed, that can't be named.

N: You are saying it is not feeling; it is similar to feeling, but it is not fixed?

DB: Yes. I am just considering that that could exist if we say that there is no thought. I am trying to clarify this.

JK: Yes, there is no thought.

DB: What does that really mean?

JK: What it really means is thought is movement, thought is time. Right? In that emptiness there is no time or thought.

DB: Yes, and perhaps no sense of the existence of an entity inside.

JK: Absolutely, of course. The existence of the entity is the bundle of memories, the past.

DB: But that existence is not only thought thinking about it, but also the feeling that it is there; you get a sort of feeling inside.

JK: A feeling, yes. There is no being. There is nothing. If there is a feeling of the being continuing . . .

DB: Yes, even though it doesn't seem possible to verbalize this. . . . It would be a state without desire. How can we know if this state is real, is genuine?

JK: That is what I am asking. How do we know or realize that this is so? In other words, do you want proof of it?

N: Not proof, but communication of that state.

JK: Now, wait a minute. Suppose someone has this peculiar compassion. How can he communicate it to me? If I am living in pleasure and all that, he can't!

N: No, but I am prepared to listen to him.

JK: Prepared to listen, but how deeply? You will go as long as it is safe, secure.

N: No, not necessarily.

JK: The man says there is no being. And one's whole life has been this becoming, being, and so on. And, in that state, he says there is no being at all. In other words, there is no "me." Right? Now, you say, "Show it to me." It can be shown only through certain qualities that it has, certain actions. What are the actions of a mind that is totally empty of being? Actions at what level? Actions in the physical world?

N: Partly.

JK: Mostly that. All right, this man has got this sense of emptiness, and there is no being. He is not acting from self-centred interest. His actions are in the world of daily living, and you can judge whether he is a hypocrite, whether he says something and contradicts it the next moment, or whether he is actually living this compassion and not just saying, "I feel compassionate."

DB: But if one is not doing the same, one can't tell.

JK: That's right. That is what I am saying.

N: We can't judge him.

JK: You can't. So how can he convey to us in words that peculiar quality of mind? He can describe, go round it, but he can't give the essence of it. Dr. Bohm, for example, could discuss with Einstein; they were on the same level. And he and I can discuss up to a certain point. If one has this sense of not being, of emptiness, the other can go very close but can never enter or come upon that mind unless he has it!

N: Is there any way of communicating, for one who is open, but not through words?

JK: We have said compassion. It is not, as David pointed out just now, "I feel compassionate." That is altogether wrong. [*Pause*] You see, after all in daily life such a mind acts without the "me," without the "ego." And therefore it might make a mistake but corrects it immediately; it is not carrying that mistake.

N: It is not stuck.

JK: Not stuck. But we must be very careful here not to find an excuse for wrong [*laughs*]!

So we come to that point that we discussed earlier. What then is meditation? Right? For the man who is becoming or being, meditation has no meaning whatsoever. That is a tremendous statement. Now, when there is this not-being-or-becoming, what is meditation? It must be totally unconscious, totally uninvited.

DB: Do you mean without conscious intention?

JK: Yes, I think that is right. Would you say—I hope this doesn't sound silly—that the universe, the cosmic order, is in meditation?

DB: Well, if it is alive, then we would have to look at it that way.

JK: No, no. It is in a state of meditation.

DB: Yes.

JK: I think that is right. I stick to that.

DB: We should try to go further into what is meditation. What is it doing?

N: If you say that the universe is in meditation, is the expression of it order? What order can we discern which would indicate cosmic or universal meditation?

J K: The sunrise and sunset; all the stars, the planets are order. The whole thing is in perfect order.

D B: We have to connect this with meditation. According to the dictionary, the meaning of meditation is to reflect, to turn something over in the mind, and to pay close attention.

J K: And also to measure.

D B: That is a further meaning, but it is to weigh, to ponder; it means measure, in the sense of weighing.

J K: Weighing—that's it. Ponder, think over, and so on.

D B: To weigh the significance of something. Now, is that what you mean?

J K: No.

D B: Then why do you use the word?

N: I am told that, in English, "contemplation" has a different connotation from "meditation." "Contemplation" implies a deeper state of mind.

D B: It is hard to know. The word "contemplate" comes from the word "temple," really.

J K: Yes, that's right.

D B: Its basic meaning is to create an open space.

J K: Is that an open space between God and me?

D B: That is the way the word arose.

J K: Quite.

N: The Sanskrit word "dhyana" doesn't have the same connotation as meditation.

JK: No.

N: Because meditation has the overtones of measurement, and probably, in an oblique way, that measurement is order.

JK: No, I don't want to bring in order; let's leave the word "order" out. We have been through that and beaten it to death!

DB: Why do you use the word "meditation"?

JK: Don't let's use it.

DB: Let's find out what you really mean here.

JK: Would you say a state of infinity? A measureless state?

DB: Yes.

JK: There is no division of any kind. You see, we are giving lots of descriptions, but it is not that.

DB: Yes, but is there any sense of the mind being in some way aware of itself? Is that what you are trying to say? At other times you have said that the mind is emptying itself of content.

JK: What are you trying to get at?

DB: I am asking whether it is not merely infinite, but if something more is involved.

JK: Oh, much more.

DB: We said that this content is the past, which has become disorder. Then you could say that this emptying of content in some sense is constantly cleaning up the past. Would you agree to that?

JK: No, no.

DB: When you say the mind is emptying itself of content . . .

JK: *Has* emptied itself.

DB: All right then. When the past is cleaned up, then you say that is meditation.

JK: That is meditation, not contemplation . . . of what?

N: This is just a beginning.

JK: Beginning?

N: The emptying of the past.

JK: That emptying of the past, which is anger, jealousy, beliefs, dogmas, attachments, and so on, must be done. If that is not emptied, if any part of that exists, it will inevitably lead to illusion. The brain or the mind must be totally free of all illusion, illusion brought by desire, by hope, by wanting security, and all that.

DB: Are you saying that when this is done, it opens the door to something broader, deeper?

JK: Yes. Otherwise life has no meaning; it is just repeating this pattern.

N: What exactly did you mean when you said that the universe is meditation?

JK: I feel that way, yes. Meditation is a state of "non-movement movement."

DB: Could we say first of all that the universe is not actually governed by its past? You see, the universe creates certain forms which are relatively constant, so that people who look at it superficially only see that, and it seems then to be determined from the past.

JK: Yes, but it is not governed by the past. It is creative, moving.

DB: And then this movement is order.

JK: Would you, as a scientist, accept such a thing?

DB: Well, as a matter of fact I would [*laughs*]!

JK: Are we both crazy? Let's put the question another way: Is it really possible for time to end—time being the past, the whole idea of time—to have no tomorrow at all? Of course there is tomorrow; you have to go to a talk in the morning, and I have to, and so on. But the feeling, the actual reality of having no tomorrow—I think that is the healthiest way of living. Which doesn't mean that I become irresponsible! That would be too childish.

DB: It is merely a question of physical time, which is a certain part of natural order.

JK: Of course. That is understood.

DB: The question is whether we have a sense of experiencing past and future or whether we are free of that sense.

JK: I am asking you as a scientist, is the universe based on time?

DB: I would say no, but you see the general way ...

JK: That is all I want. You say no! And can the brain, which has evolved in time ... ?

DB: Well, has it evolved in time? Rather it has become entangled in time. Because the brain is part of the universe, which we say is not based on time.

JK: I agree.

DB: Thought has entangled the brain in time.

JK: All right. Can that entanglement be unravelled, freed, so that the universe is the mind? You follow? If the universe is not of time, can the mind, which has been entangled in time, unravel itself and so be the universe? You follow what I am trying to say?

DB: Yes.

JK: That is order.

DB: That is order. Now, would you say that is meditation?

JK: That is it. I would call that meditation, not in the ordinary dictionary sense of pondering and all that, but a state of meditation in which there is no element of the past.

DB: You say the mind is disentangling itself from time and also really disentangling the brain from time?

JK: Yes. Would you accept that?

DB: Yes.

JK: As a theory?

DB: Yes, as a proposal.

JK: No, I don't want it as a proposal.

DB: What do you mean by theory?

JK: Theory—when somebody comes along and says this is real meditation.

DB: All right.

JK: Wait. Somebody says one can live this way; life has an extraordinary meaning in it, full of compassion, and that every act in the physical world can be corrected immediately, and so on. Would you, as a scientist, accept such a state, or say that the man who talks of it is cuckoo?

DB: No, I wouldn't say that. I feel it is perfectly possible; it is quite compatible with anything that I know about nature.

JK: Oh, then that's all right. So one is not an unbalanced cuckoo!

DB: Part of the entanglement is that science itself has put time into a fundamental position, which helps to entangle it still further.

JK: Of course putting all this into words is not the thing, right? That is understood. But can it be communicated to another? Now, can some of us get to this, so that we can communicate it, actually?

Eleven

The Liberation of Insight

JIDDU KRISHNAMURTI: We have asked what is the origin of all human movement. Is there an original source, a ground from which all this—nature, man, the whole universe—sprang? Is it bound by time? Is it in itself complete order, beyond which there is nothing more?

And we have talked about order, whether the universe is based on time at all, and whether man can ever comprehend and live in that supreme order. We want to investigate, not merely intellectually but also profoundly, how to comprehend and live, move from that ground, that ground that is timeless, and beyond which there is nothing. Can we go on from there?

I don't know if, as a scientist, you will agree that there is such a ground, or that man can ever comprehend it, live in it; not in the sense that he is living in it, but that itself is living? Can we as human beings come to that?

DAVID BOHM: I don't know if science as it is now constituted can say much about that.

JK: Science doesn't talk about it, but would you, as a scientist, give your mind to the investigation of that?

DB: Yes, I think that implicitly science has always been concerned with trying to come to this ground, but to attempt it by studying matter to the greatest possible depth, of course, is not enough.

JK: Didn't we ask if a human being, living in this world that is in such turmoil, can be in absolute order first, as the universe is in absolute order, and comprehend an order which is universal?

DB: Yes.

JK: I can have order in myself, by careful observation, self-study, self-investigation, and understanding the nature of disorder. The very insight of that understanding dispels disorder. That's one level of order.

DB: Yes, that's the level that most of us have been concerned with till now, you see. We see this disorder going on in the world and in ourselves, and we say it is necessary to be aware of it and to observe it and, as you say, to dispel it.

JK: But that's a very small affair.

DB: Yes, but we agreed that people generally don't feel it is a small affair. They feel that clearing up the disorder in themselves and the world would be a very big thing, and perhaps all that is necessary.

JK: But I am referring to the fairly intelligent, knowledgeable, and cultured human being—"cultured" meaning civilized. He can, with a great deal of enquiry and investigation, come to the point when he can bring order in himself.

DB: Then some people would say, if only we could bring that order into the whole of society.

JK: Well, we will, if human beings are all tremendously orderly in that inward sense, perhaps create a new society. But that again is a very small affair.

DB: I understand that, but I feel we should go into it carefully because people commonly don't see it as small. Only a few have seen that there's something beyond that.

JK: Much more beyond that.

DB: Perhaps it might be worth thinking about why it is not enough to go into the order of man and society, just to produce orderly living. In what sense is that not enough?

JK: Because we live in chaos. We think that to bring order is a tremendous affair, but in itself it isn't. I can put my room in order, so that it gives me certain space, certain freedom; I know where things are, I can go directly to them. That's a physical order. Can I put things in myself in order, which means not to have conflict, not to have comparison, not to have any sense of "me" and "you" and "they," everything that brings about such division, out of which grows conflict? That's simple. If I'm a Hindu and you are a Muslim, we are eternally at war with each other.

DB: Yes, and in every community people fall apart in the same way.

JK: All society breaks up that way, but if one understands that, and profoundly realizes it, it's finished.

DB: Suppose we say we have achieved that. Then what? I think some people might feel it's so far away that it doesn't interest them. They might say, wait till we achieve it before we worry about the other.

JK: All right, sir, let's start again. I'm in disorder, physically and psychologically. Around me the society in which I live is also utterly confused. There is a great deal of injustice; it is a miserable affair. I can see that very simply. I can see that my generation and past generations have contributed to this. And I can do something about it. That's simple. I can say, well, I'll put my house in order. The house is myself and it must be in order before I can move further.

DB: But suppose somebody says, "My house is not in order"?

JK: All right, my house is in disorder. Then let me put it in order, which is fairly simple. If I apply my mind and my heart to the resolution of that, it's fairly clear. But we don't want to do that. We find it tremendously difficult because we are so bound to the past, to our habits and our attitudes. We don't seem to have the energy, the courage, the vitality, to move out of it.

DB: What's not simple is to know what will produce that energy and courage. What will change all this?

JK: I think that what will change all this is to have an insight into it.

DB: The key point seems to be that without insight, nothing can change.

JK: Will insight really alter the whole structure and nature of my being? That is the question, isn't it?

DB: What seems to be implied is that if we look at a rather small question like the order of daily life, it will not involve our whole being.

JK: No, of course not.

DB: And therefore the insight will be inadequate.

JK: Yes, it's like being tied to something, to a belief, to a person, an idea, some habit, some experience. That inevitably must create disorder, because being tied implies dependence, the escape from one's own loneliness, fear. Now, total insight into that attachment clears it all away.

DB: Yes. I think we are saying that the self is a centre creating darkness or clouds in the mind, and insight penetrates that. It could dispel the clouds so that there would be clarity and the problem would vanish.

JK: That's right, vanish.

DB: But that would take a very intense, total insight.

JK: That's right, but are we willing to go through that? Or is our attachment or tie to something so strong that we're unwilling to let go? That is the case with most people. Unfortunately, it's only the very few who want to do this kind of thing.

Now, can insight wipe away, banish, dissolve the whole movement of being tied, attached, dependent, lonely, with one blow as it were? I think it can. I think it happens when there is profound insight. That insight is not merely the movement of memory, knowledge, experience; it is totally different from all that.

DB: It is insight into the whole of disorder, into the source of all disorder of a psychological nature.

JK: It is all that.

DB: With that insight, the mind can clear up, and then it would be possible to approach the cosmic order.

JK: That's what I want to get at. That's much more interesting than this. Any serious man must put his house in order. And that must be complete order, order in the whole of man, not order in a particular direction. The particular resolution of a particular problem is not the resolution of the whole.

DB: The key point is that finding the source, the root that generates the whole, is the only way.

JK: Yes, that's right.

DB: Because if we try to deal with a particular problem, it's still always coming from the source.

JK: The source is the "me." That little source, little pond, little stream, apart from the great source must dry up.

DB: Yes, the little stream confuses itself with the great one, I think.

JK: Yes, we're not talking about the great stream, the immense movement of life; we're talking about the little "me" with the little movement, little apprehensions, and so on, that is creating disorder. As long as that centre, which is the very essence of disorder, is not dissolved, there is no order.

So at that level it is clear. Can we go on from there? Now, is there another order totally different from this? This is man-made disorder, and therefore man-made order. The human mind, realizing that and seeing the disorder that it can bring about in itself, then begins to ask if there is an order that is totally different, of a dimension which it is necessary to find, because this man-made order is such a small affair.

I can put my house in order. All right. Then what? Perhaps if many of us do it, we'll have a better society. That is admitted, relevant, necessary, but it has its limitation. Now, a human being who has really deeply understood the disorder made by human beings and its effect on society asks, "Is there an order that is beyond all this?" The human mind isn't satisfied with merely physical order. That has limitations, boundaries, so he says, "I've understood that; let's move."

DB: How do we get into that question? Even in science, men seek the order of the universe looking to the end or the beginning or to the depth of its structure. Many have sought the absolute, and the word "absolute" means free of all limitation, all dependence, all imperfection. The "absolute" has been the source of tremendous illusion, of course, because the limited self seeks to capture the absolute.

JK: Of course, that's impossible. So how do we approach this? How do we answer this question? As a scientist, would you say there is an order which is beyond all human order and disorder?

DB: Science is not able to say anything because any order discovered by science is relative. Not knowing what to do, men have felt the need for the absolute, and not knowing how to get it, they have created the illusion of it in religion and in science or in many other ways.

JK: So what shall I do? As a human being who is the totality of human beings, there is order in my life. That order is naturally brought about through insight, and so perhaps it will affect society. Move from that. The enquiry then is, is there an order that is not man-made? Let's put it that way. I won't even call it absolute order.

Man has sought a different dimension and perhaps used the word "order." He has sought a different dimension, because he has understood this dimension. He has lived in it, he has suffered in it, he has gone through all kinds of mess and misery, and he has come to the end of all that. Not just verbally, but he has actually come to the end of all that. You may say there are very few people who do that, but this question must be put.

DB: Would this question have any significance to a person who had not come to the end of that?

JK: I think it would. Because even if only intellectually he may see the limitations of that.

DB: Yes, it is important for him to see this even before he has finished with it.

JK: How does the mind approach this problem? I think man has struggled to find this out, sir. All so-called religious people—the mystics, the saints, with their illusions—have attempted to grasp this. They have tried to understand something which is not all this. Does it come about through, if I may use the word, "meditation," as measure?

DB: The original meaning of the word "meditation" is to measure, to ponder, to weigh the value and significance. Perhaps that may have meant that such a measurement would only have significance for seeing that there is disorder.

JK: That's what I would say, that measurement can exist only where there is disorder. We are using the word "meditation" not as "measure" or even to "ponder" or "think over," but as meditation that is the outcome of bringing order in the house and moving from there.

DB: So if we see things are in disorder in the mind, then what is meditation?

JK: First the mind must be free of measurement. Otherwise it can't enter into the other. All *effort* to bring order into disorder is disorder.

DB: So we are saying that it is the attempt to control that is wrong; we see that it has no meaning. And now we say there's no control. What do we do?

JK: No, no, no. If I have an insight into the whole nature of control, which is measure, that liberates the mind from that burden.

DB: Yes. Could you explain the nature of this insight, what it means?

JK: Insight is not a movement from knowledge, thought, remembrance, but the cessation of all that, to look at the problem with pure observation, without any pressure, without any motive, to observe the whole movement of measurement.

DB: Yes, I think we can see that measurement is the same as becoming and the attempt of the mind to measure itself, to control itself, to set itself a goal, is the very source of the disorder.

JK: That is the very source of disorder.

DB: In a way that was the wrong way of looking at it, a wrong turning, when man extended measurement from the external sphere into the mind.

JK: Yes.

DB: But now the first reaction would be that if we don't control this thing, it will go wild. That's what someone might fear.

JK: Yes. But, you see, if I have an insight into measurement, that very insight not only banishes all movement of measurement, but there is a different order. It doesn't go wild; on the contrary.

DB: It does not go wild because it has begun in order. It is really the attempt to measure it that makes it go wild.

JK: Yes, that's it. The measurement becomes wild; it is confusion.

Now, let's proceed. After establishing all this, can the mind, through meditation—using the word "meditation" without any sense of measurement, comparison—find an order, a state where there is something that is not man-made? I've been through all the man-made things and they are all limited. There is no freedom in them; there is chaos.

DB: When you say you've been through man-made things, what are they?

JK: Religion, worship, prayers, science, anxieties, sorrow, attachment, detachment, loneliness, suffering, confusion, ache—all that.

DB: Also all the attempts by revolution.

JK: Of course, physical revolution, psychological revolution. Those are all man-made. And also so many people have put this question, and then they say "God." That is another concept, and that very concept creates disorder.

Now, one has finished with all that. Then the question is, is there something beyond all this that is never touched by human thought, mind?

DB: Yes. Now, that makes a difficult point: not touched by the human mind, but mind might go beyond thought.

JK: Yes, that's what I want.

DB: Do you mean by the mind only thought, feeling, desire, will, or something much more?

JK: For the time being we have said the human mind is all that.

DB: But it's not; the mind is now considered to be limited.

JK: No. As long as the human mind is caught in that, it is limited.

DB: Yes, the human mind has potential.

JK: Tremendous potential.

DB: Which it does not realize now when it is caught in thought, feeling, desire, will, and that sort of thing.

JK: That's right.

DB: Then we'll say that which is beyond this is not touched by this limited sort of mind.

JK: Yes.
   [*Pause*]

DB: Now, what will we mean by the mind which is beyond this limit?

JK: First of all, sir, is there such a mind? Is there such a mind that actually, not theoretically or romantically, can say, "I've been through this"?

DB: You mean through the limited stuff.

JK: Yes. And being through it means finished with it. Is there such a mind? Or does it only think it has finished with it, and therefore it creates the illusion that there is something else? I won't accept that. A human being, a person, X, says, "I have understood this. I have seen the limitation of all this. I have been through it, and I have come to the end of it. And this mind, having come to the end of it, is no longer the limited mind." Is there a mind which is totally limitless?

DB: What is the relation between that unlimited mind and the brain?

JK: I want to be clear on this point. This mind, brain, the whole of it, the whole nature and the structure of the mind, includes the emotions, the brain, the reactions, physical responses—all that. This mind has lived in turmoil, in chaos, in loneliness, and it has understood all that, has had a profound insight into it. Having such a deep insight cleared the field. This mind is no longer that mind.

DB: Yes, it's no longer the original limited mind.

JK: Yes. Not only that, no longer the limited mind, the damaged mind. Damaged mind means damaged emotions, damaged brain.

DB: The cells themselves are not in the right order.

JK: Quite. But when there is this insight and therefore order, the damage is undone.

DB: By reasoning you can see it's quite possible, because you can say the damage was done by disorderly thoughts and feelings, which overexcite the cells and disrupt them, and now with the insight, that stops and there is a new process.

JK: Yes, it's like a person going for fifty years in a certain direction. If he realizes suddenly that it's the wrong direction, the whole brain changes.

DB: It changes at the core, and then the wrong structure is dismantled and healed. That may take time.

JK: That's right.

DB: But the insight . . .

JK: Is the factor that changes.

DB: And that insight does not take time.

JK: That's right.

DB: But it means that the whole process has changed the origin.

JK: The limited mind with all its consciousness and its content says it is over. Now, is that mind—which has been limited but has had insight into its limitation and moved away from that limitation—an actuality? Is it then something that is really tremendously revolutionary? And therefore it is no longer the human mind?

When the human mind with its consciousness, which is limited, is ended, then what is the mind?

DB: Yes, and what is the person, what is the human being?

JK: What is a human being, then? And then what is the relationship between that mind, which is not man-made, and the man-made mind? Can one observe, really, deeply, without any prejudice, whether such a mind exists? Can the mind, conditioned by man, uncondition itself so completely that it's no longer man-made? Can the man-made mind liberate itself completely from itself?

DB: Yes, of course that's a somewhat paradoxical statement.

JK: Of course it's paradoxical; but it's actual, it is so. Let's begin again. One can observe that the consciousness of humanity is its content. And its content is all the man-made things—anxiety, fear, and so on. And it is not only the particular, it is the general. Having had an insight into this, it has cleansed itself from that.

DB: Well, that implies that it was always potentially more than that, but insight enabled it to be free of that. Is that what you mean?

JK: I won't say that insight is potential.

DB: There is a little difficulty of language if you say the brain or the mind had an insight into its own conditioning, and then you're almost saying it became something else.

JK: Yes, I am saying that, I am saying that. The insight transforms the man-made mind.

DB: Yes, but then it's no longer the man-made mind.

JK: It's no longer. That insight means the wiping away of all the content of consciousness. Not bit by bit by bit—the totality of it. And that insight is not the result of man's endeavour.

DB: Yes, but then that seems to raise the question of where it comes from.

JK: All right. Where does it come from? Yes. In the brain itself, in the mind itself.

DB: Which, the brain or the mind?

JK: Mind—I'm saying the whole of it.

DB: We say there is mind, right?

JK: Just a minute, sir. Let's go slowly. It's rather interesting. Consciousness is man-made, general and particular. And logically, reasonably one sees the limitations of it. Then the mind has gone much further. Then it comes to a point when it asks, "Can all this be wiped away at one breath, one blow, one movement?" And that movement is insight, the movement of insight. It is still in the mind, but it's not born of that consciousness.

DB: Yes. Then you are saying the mind has the possibility, the potential, of moving beyond that consciousness.

JK: Yes.

DB: The brain, mind can do that, but it hasn't generally done it.

JK: Yes. Now, having done all this, is there a mind which is not only not man-made but that man cannot conceive, cannot create, that is not an illusion? Is there such a mind?

DB: Well, I think what you are saying is this mind, having freed itself from the general and particular structure of the consciousness of mankind, from its limits, is now much greater. Now, you say that this mind is raising a question.

JK: This mind is raising the question.

DB: Which is what?

JK: Which is, first, is that mind free from the man-made mind? That's the first question.

DB: It may be an illusion.

JK: Illusion is what I want to get at. One has to be very clear. No, it is not an illusion, because he sees measurement is an illusion; he knows the nature of illusion; that it is born of desire. And illusions must create limitation and so on. He has not only understood it, he's over it.

DB: He's free of desire.

JK: Free of desire. That is his nature. I don't want to put it so brutally. Free of desire.

DB: It is full of energy.

JK: Yes. So this mind, which is no longer general and particular, is therefore not limited; the limitation has been broken down through insight, and therefore the mind is no longer that conditioned mind. Now, then, what is that mind? Being aware that it is no longer caught in illusion?

DB: Yes; but we were saying it raised a question about whether there is something much greater.

JK: Yes, that's why I'm raising the question. Is there a mind which is not man-made? And if there is, what is its relationship to the man-made mind?

You see, every form of assertion, every form of verbal statement is not that. So we're asking if there is a mind which is not man-made. I think that can only be asked when the limitations are ended; otherwise it's just a foolish question.

So one must be absolutely free of all this. Then only can you put that question. Then you put that question—not "you"—then the question is raised: Is there a mind that is not man-made, and if there is such a mind, what is its relationship to the man-made mind? Now, is there such a mind? Of course there is. Of course, sir. Without being dogmatic or personal, or all that business, there is. But it is not God; we've been through all that.

There is. Then the next question is what is the relationship of that to the human mind, man-made mind? Has it any relationship? Has this relationship to that? Obviously not. The man-made mind has no relationship to that. But that has a relationship to this.

DB: Yes, but not to the illusions in the man-made mind.

JK: Let's be clear. My mind is the human mind. It has illusions, desires, and so on. And there is that other mind which has not, which is beyond all limitations. This illusory mind, the man-made mind, is always seeking that.

DB: Yes, that's its main trouble.

JK: Yes, that's its main trouble. It is measuring, it is "progressing," saying "I am getting nearer, going farther." And this mind, the human mind, the mind that's made by human beings, the man-made mind, is always seeking that, and therefore it's creating more and more mischief, confusion. This man-made mind has no relationship to that.

Now, has that any relationship to this?

DB: I was suggesting that it would have to have, but that if we take the illusions which are in the mind, such as desire and fear and so on, it has no relationship to those, because they are figments anyway.

JK: Yes, understood.

DB: But that can have a relationship to the man-made mind in understanding its true structure.

JK: Are you saying, sir, that that mind has a relationship to the human mind the moment it's moving away from the limitations?

DB: Yes, but in understanding those limitations it moves away.

JK: Yes, moves away. Then that has a relationship.

DB: We have to get the words right. The mind that is not limited, that is not man-made, cannot be related to the illusions that are in the man-made mind.

JK: No. Agreed.

DB: But it has to be related to the source, as it were, to the real nature of the man-made mind, which is behind the illusion.

JK: The man-made mind is based on what?

DB: Well, on all these things we have said.

JK: Yes, which is its nature. Therefore, how can that have a relationship to this, even basically?

DB: The only relationship is in understanding it, so that some communication would be possible, which might communicate to the other person.

JK: No, I'm questioning that.

DB: Yes, because you were saying that the mind that is not man-made may be related to the limited mind and not the other way round.

JK: I even question that.

DB: Yes, all right. You are changing that.

JK: No, I'm just pushing it a little.

DB: It may or may not be so. Is that what you mean by questioning it?

JK: Yes, I'm questioning it. What is the relationship then of love to jealousy? It has none.

DB: Not to jealousy itself, no; that is an illusion, but . . .

JK: I'm taking two words, say "love" and "hatred." Love and hatred really have no relationship to each other.

DB: No, not really. I think that love might understand the origin of hatred, you see.

JK: Ah, yes, yes. I see. You're saying that love can understand the origin of hatred and how hatred arises. Does love understand that?

DB: Well, I think that in some sense it understands its origin in the man-made mind, and that having seen the man-made mind and all its structure and moved away . . .

JK: Are we saying, sir, that love—using that word for the moment—has a relationship to non-love?

DB: Only in the sense of dissolving it.

JK: I'm not sure, I'm not sure. We must be awfully careful here. Or the ending of itself . . .

DB: Which is it?

JK: With the ending of hatred, the other is; not the other has relationship to the understanding of hatred.

DB: We have to ask how it gets started then, you see.

JK: Suppose I have hatred. I can see the origin of it: It's because you insulted me.

DB: That's a superficial notion of the origin. Why one behaves so irrationally is the deeper origin. There's nothing real if you merely insult me, so why should I respond to the insult?

JK: Because all my conditioning is that.

DB: Yes, that's what I mean by your understanding the origin.

JK: But does love help me to understand the origin of hatred?

DB: No, but I think that someone in hatred, moving, understands the origin and moves away.

JK: Moving away, then the other is. The other cannot help the movement away.

DB: No, but suppose one human being has this love and another has not. Can the first one communicate something which will start the movement in the second one?

JK: That means A can influence B.

DB: Not influence, but, for example, why should anybody be talking about any of this?

JK: That's a different matter. No, the question, sir, is, is hate dispelled by love?

DB: No.

JK: Or, in the understanding of hatred and the ending of it, the other is?

DB: That's right. But say that here in A the other now is, that A has reached that. Love is for A, and he sees B, and we're asking what he is going to do. You see, that's the question. What is he going to do?

JK: What is the relationship between the two? My wife loves, and I hate. She can talk to me, she can point out to me my unreasonableness, and so on, but her love is not going to transform the source of my hatred.

DB: That's clear, yes, except love is the energy which will be behind the talk.

JK: Behind the talk, yes.

DB: The love itself doesn't sort of go in there and dissolve the hate.

JK: Of course not. That's romantic.

So the man who hates, who has an insight into the source of it, the cause of it, the movement of it, and ends it, has the other.

DB: Yes. We say A is the one who has seen all this, and he now has the energy to put it to B. It's up to B what happens.

JK: Of course. I think we had better pursue this.

Twelve

The Intelligence of Love

16 SEPTEMBER 1980, BROCKWOOD PARK, HAMPSHIRE

JIDDU KRISHNAMURTI: We have been saying that a human being who has worked his way through all the problems of life, both physical and psychological, and has really grasped the full significance of freedom from psychological memories and conflicts and travails, comes to a point where the mind finds itself free but hasn't gathered that supreme energy needed to go beyond itself.

Can the mind, brain, the whole psychological structure, ever be free from all conflict, from all shadow of any disturbance? Or is the idea of complete freedom an illusion?

DAVID BOHM: That's one possibility. Then some people would say we could have partial freedom.

JK: Or is the human condition so determined by the past, by its own conditioning, that it can never free itself from it, as some philosophers have stated?

There have been some deeply nonsectarian religious people, totally free from all organized religions and beliefs, rituals, dogmas, who have said it can be done, but very few have said it. Or some say it will take a very long time, that you must go through various lives and suffer all kinds of miseries and ultimately you come to that. But we are not thinking in terms of time. We are asking if a human being—granting, knowing that he is conditioned, deeply, profoundly, so that his whole being is that—can ever free himself. And if he can, what is beyond? That's what we were coming to.

Would that question be reasonable or valid unless the mind has really finished with all the travail of life? We said our minds are man-made and asked if there is a mind that is not man-made. Is it possible that the man-made mind can free itself from its own man-made mechanical mind? How shall we find this out?

DB: There's a difficult thing to express here. If this mind is totally man-made, totally conditioned, then in what sense can it get out of it? If you say that it had at least the possibility of something beyond . . .

JK: Then it becomes a reward, a temptation.

DB: Logically it may appear to be inconsistent to say that the mind is totally conditioned and yet it's going to get out.

JK: If one admits that there is a part that is not conditioned, then we enter into quite another thing.

DB: That may be another inconsistency.

JK: Yes. We have been saying that the mind, although deeply conditioned, can free itself through insight. That is the real clue to this. Would you agree to that?

DB: Yes.

JK: We went into what the nature of insight is. Can insight un-condition the mind completely and wipe away completely all the illusions, all the desires? Or is it partial?

DB: If we say the mind is totally conditioned, it suggests something static, which would never change. Now, if we say the mind is always in movement, then it seems in some way it becomes impossible to say what it is at this moment. We couldn't say it has been totally conditioned.

JK: No, let's say I'm totally conditioned; it's in movement, but the movement is within a border, within a certain field. And the field is very definitely marked out. The mind can expand it and contract it, but the boundary is very, very limited, definite. Now, it is always moving within that limitation. Can it die away from that?

DB: That's the point; that's another kind of movement. It's kind of in another dimension, I think you've said.

JK: Yes. And we say it is possible through insight, which is also a movement, a totally different kind of movement.

DB: Yes, but then we say *that* movement does not originate in the individual nor in the general mind.

JK: Yes. It is not the insight of the particular or the general. We are then stating something quite outrageous.

DB: That rather violates most of the sort of logic that people use. Either the particular or the general should cover everything, in terms of ordinary logic.

JK: Yes.

DB: Now, if you're saying there's something beyond both, this is already a question which has not been stated, and I think it has great importance.

JK: How do we then state it, or how do we then come to it?

DB: People divide themselves roughly into two groups. One group feels the ground is the concrete, particular daily activity. The other group feels that the general, the universal, is the ground. One is the more practical type, and the other the more philosophical type. In general, this division has been visible throughout history and also in everyday life, wherever you look.

JK: But, sir, is the general separate from the particular?

DB: It's not. Most people agree with that, but people tend to give emphasis to one or the other. Some give emphasis to the particular, saying the general is there but if you take care of the particular the general will be all right. The others say the general is the main thing, the universal, and by getting that right you'll get the particular right. So there's been a kind of imbalance to one side or the other, a bias in the mind of man. What's being raised here is the notion that it is neither the general nor the particular.

JK: That's right. That's just it. Can we discuss it logically? Using your expertise, your scientific brain and this man who is not all that, can we have a conversation to find out if the general and particular are one, not divided at all?

So where are we now? We are neither the particular nor the general. That's a statement that can hardly be accepted reasonably.

DB: Well, it's reasonable if you take thought to be a movement rather than a content. Then thought is the movement between the particular and the general.

JK: Thought is a movement. Quite; we agree to that. But thought is the general and thought is the particular.

DB: Thought is also the movement. In the movement it goes beyond being one or the other.

JK: Does it?

DB: Well, it can. Ordinarily it does not, because ordinarily thought is caught on one side or the other.

JK: That's the whole point, isn't it? Ordinarily the general and the particular are in the same area.

DB: Yes, and you fix on one or the other.

JK: Yes, but in the same area, in the same field. And thought is the movement between the two. Thought has created both.

DB: Yes, it has created both and moves between.

JK: Yes, between and around and in that area. And it has been doing this for millennia.

DB: Yes, and most people would feel that's all it can do.

JK: Yes. Now we are saying that, when thought ends, that movement which thought has created also comes to an end. Therefore time comes to an end.

DB: We should go more slowly here, because you see it's a jump from thought to time. We've gone into it before, but it's still a jump.

JK: Sorry. Right. Let's see. Thought has created the general and the particular, and thought is a movement that connects the two. Thought moves round it, so it is still in the same area.

DB: Yes, and doing that it has created time, which is part of the general and the particular. Time is a particular time and also a general time.

JK: Yes, but you see, thought is time.

DB: Well, that's another question. We have said that thought has a content which is about time, and that thought is a movement which is time. It could be said to be moving from the past into the future.

JK: But, sir, thought is based on time, thought is the outcome of time.

DB: Yes, but then does that mean that time exists beyond thought? If you say thought is based on time, then time is more fundamental than thought.

JK: Yes.

DB: We have to go into that. You could say that time is something which was there before thought, or at least is at the origin of thought.

JK: Time is there when there is the accumulation of knowledge.

DB: Well, that has come out of thought to some extent.

JK: No, I act and learn. That action is not based on previous knowledge, but I do something, and in the doing I learn.

DB: Yes. Then that learning is registered in the memory.

JK: Yes. So is not thought essentially the movement of time?

DB: We have to say in what sense this learning is the movement of time. You can say that when we learn, it is registered, and then that what you have learned operates in the next experience.

JK: Yes. The past is always moving to the present.

DB: Yes, and mixing, fusing with the present. And the two together are again registered as the next experience.

JK: So are we saying time is different from thought, or time is thought?

DB: We are saying this movement of learning, and the response of memory into experience and then registering, is time, and that is also thought.

JK: Yes, that is thought. Is there a time apart from thought?

DB: That's another question. Would we say that physically or in the cosmos time has a significance apart from thought?

JK: Physically, yes, I understand that.

DB: So then we're saying in the mind or psychologically.

JK: Psychologically. As long as there is psychological accumulation as knowledge, as the "me," and so on, there is time. It is based on time.

DB: Wherever there is accumulation, there is time.

JK: Yes, that's the point. Wherever there is accumulation, there is time.

DB: Which turns the thing around, because usually you say time is first and in time you accumulate.

JK: No, I would put it round the other way, personally.

DB: Yes. But it's important to see that it's put the other way. Then we'd say, suppose there is no accumulation? Then what?

JK: Then—that's the whole point—there is no time. As long as I am accumulating, gathering, becoming, there is the process of time. But if there is no gathering, no becoming, no accumulation, where does psychological time exist? So thought is the outcome of psychological accumulation, and that accumulation, that gathering, gives it a sense of continuity, which is time.

DB: It seems it's in movement. Whatever has been accumulated is responding to the present, with the projection of the future, and then that is again registered. Now, the accumulation of all that's registered is in the order of time—one time, the next time, and so on.

JK: That's right. So we're saying thought is time. Psychological accumulation is thought and time.

DB: We're saying that we happen to have two words when really we only need one.

JK: One word. That's right.

DB: Because we have two words, we look for two things.

JK: Yes. There is only one movement, which is time and thought, time plus thought, or time/thought. Now, can the mind, which has moved for millennia in that area, free itself from that?

DB: Yes. Now, why is the mind bound up? Let's see exactly what's holding the mind.

JK: Accumulation.

DB: Yes, but why does the mind continue to accumulate?

JK: Because in accumulation there is apparent safety, there is apparent security.

DB: The accumulation of physical food may provide a certain kind of security. And then since no distinction was made between the outer and the inner, there was the feeling that one could accumulate inwardly either experiences or some knowledge of what to do.

JK: Are we saying the outward physical accumulation is necessary for security, and that same movement, same idea, same urge, moves into the field of the psychological, so we accumulate there, hoping to be secure?

DB: Yes, inwardly hoping to accumulate present memories or relationships or things you could count on, principles you could count on.

JK: So psychological accumulation is the illusion of safety, protection, security?

DB: Yes. It does seem that the first mistake was that man never understood the distinction between what he has to do outside and what he has to do inside.

JK: It is the same movement, outer and inner.

DB: The movement that was right outwardly man carried inward, without knowing that that would make trouble.

JK: So where are we now? A human being realizes all this, has come to the point when he says, "Can I really be free from this accumulated security and thought and psychological time?" Is that possible?

DB: Well, if we say that it had this origin, then it should be possible to dismantle it, but, if it were built into us, nothing could be done.

JK: Of course not. It is not built into us.

DB: Most people act as though they believe it was.

JK: Of course, that's absurd.

DB: If it's not built into us, then the possibility exists for us to change. Because in some way it was built up in the first place through time.

JK: If we say it is built in, then we are in a hopeless state.

DB: Yes, and I think that's one of the difficulties of people who use evolution. They're hoping by bringing in evolution to get out of this static boundary. They don't realize that evolution is the same thing, or that it's even worse, it's the very means by which the trap was made.

JK: Yes. So as a human being, I have come to that point. I realize all this. I'm fully aware of the nature of this. And my next question is can this mind move on from this field altogether and enter, perhaps, into a totally different dimension? And we said it can only happen when there is insight.

DB: It seems that insight arises when one questions this whole thing very deeply and sees it doesn't make sense.

JK: Yes. Now, having had insight into this and seen its limitation, and looking beyond it—what is there beyond?

DB: It's very difficult to even bring this into words, but we said something has to be done on this line.

JK: Yes. I think it has to be put into words.

DB: Could you say why? Because many people might feel we should leave this entirely nonverbal.

JK: Can we say the word is not the thing? Whatever the description, it is not the real, not the truth, however much you embellish or diminish it. We recognize that the word is not that, then what is there beyond all this? Can my mind be so desireless that it won't create an illusion, something beyond?

DB: Then it's a question of desire; desire must be in this time process.

JK: Desire is time. Being, becoming, is based on desire.

DB: They are one and the same, really.

JK: Yes, one and the same. Now, when one has an insight into that whole movement of desire and its capacity to create illusion, it's finished.

DB: Since this is a very crucial point, we should try to say a little more about desire: how it's intrinsic in the accumulating process, how it comes out in many different ways. For one thing you could say that as you accumulate, there comes a sense of something missing. You feel you should have more, something to complete it. Whatever you have accumulated is not complete.

JK: Yes. Could we go into the question of becoming, first? Why is it that all human beings have this urge to become? We can understand it outwardly, simply enough. Physically, you develop a muscle to make it stronger. You can find a better job, have more comfort, and so on. But why is there this need in the human mind to try to become enlightened—let's use that word for the moment—trying to become more good, better?

DB: There must be a sense of dissatisfaction with what's there already. A person feels he would like to be complete. Suppose, for example, he has accumulated memories of pleasure, but these memories are no longer adequate and he feels something more is needed.

JK: Is it dissatisfaction? Is that it?

DB: Well, wanting to get more. Eventually he feels that he must have the whole, the ultimate.

JK: I'm not at all sure whether the word "more" is not the real thorn. More: I will be more; I will have more; I will become; this whole movement of moving forward, gaining, comparing, advancing, achieving—psychologically.

DB: The word "more" is just implicit in the whole meaning of the word "accumulate." So if you're accumulating, you have to be accumulating more. There's no other way to do it.

JK: So why is there this need in the human mind?

DB: Well, we didn't see that this "more" is wrong, inwardly. If we started outwardly to use the term "more," but then carried it inward, for some reason we didn't see how destructive it was.

JK: Why? Why haven't fairly intelligent philosophers and religious people, who have spent a great part of their lives achieving, seen this very simple thing? Why haven't the intellectuals seen the simple fact that where there is accumulation there must be more?

DB: They've seen that but they don't see any harm in it.

JK: I'm not sure they see it.

DB: They are trying to get more, so they say, "We are trying to have a better life." For example, the nineteenth century was the "century of progress." Men were improving all the time.

JK: Outward progress.

DB: But they felt that man would be improving himself inwardly too.

JK: But why haven't they ever questioned this?

DB: What would make them question it?

JK: This constant struggle for the more.

DB: They thought that was necessary for progress.

JK: But is that progress? Has that same outward urge to be better moved into the psychological realm?

DB: Can we make it clear why it does harm in the psychological realm?

JK: Let's think it out. What is the harm in accumulating, psychologically? Oh, yes, it divides.

DB: What does it divide?

JK: The very nature of accumulation brings about a division between you and me and so on.

DB: Could we make that clear, because it is a crucial point? I can see that you are accumulating in your way and I accumulate in my way. Then we try to impose a common way of accumulating and that's conflict. They say everybody should be "more."

JK: That is impossible. That never takes place. I have accumulated psychologically as a Hindu; another has accumulated as a Muslim. There are thousands of divisions. Therefore accumulation in its very nature divides people, and therefore creates conflict.

So can we say, then, that in accumulation man has sought psychological security, and that security with its accumulation is the factor of human division psychologically? That's why human beings have accumulated, not realizing its consequences. Realizing that, is it possible not to accumulate?

Suppose my mind is filled with this process of accumulation, which is psychological knowledge. Can all that end? Of course it can.

DB: If the mind can get to the root of it.

JK: Of course it can. It sees that it is an illusion that in accumulation there is security.

DB: But we are saying that desire is what keeps people going on with it.

JK: Not only desire but this deep-rooted instinct to accumulate, for the future, for safety. That and desire go together. So desire plus accumulation is the factor of division, conflict. Now, I'm asking, can that end? If it ends through an action of will, it is still the same thing.

DB: That's part of desire.

JK: Yes. If it ends because of punishment or reward, it's still the same thing. So the mind, one's mind, sees this and puts all that aside. But does the mind become free of accumulation? Yes, sir, I think it can; that is, have no psychological knowledge as accumulation at all.

DB: Yes, I think that we have to consider that knowledge goes very much further than is ordinarily meant. For example, if you're getting knowledge of a microphone, you build up an image, a picture of the microphone and everything goes into that and one expects it to continue. So if you have knowledge of yourself, it builds up a picture of yourself.

JK: Can one have knowledge of oneself?

DB: No, but if you think you have, if you think that there is knowledge about what sort of person you are, that builds up into a picture, with the expectations.

JK: But, after all, if you have knowledge of yourself, you have built an image already. But once you realize psychological accumulation as knowledge is an illusion and destructive and causes infinite pain and misery, it is finished.

DB: I know certain things in knowledge, and that it's foolish to have that kind of knowledge about myself, but then there may be other kinds of knowledge which I don't recognize as knowledge.

JK: What kind, what other kinds of knowledge does one have? Preferences, likes and dislikes, prejudices, habits—all that is in the image that one has created.

DB: Yes. Now, man has developed in such a way that that image seems extraordinarily real, and therefore its qualities don't seem to be knowledge.

JK: All right, sir. So we have said accumulation is time and accumulation is security, and where there is psychological accumulation there must be division. And thought is the movement between the particular and the general, and thought is also born out of the image of what has been accumulated. All that is one's inward state. That is deeply embedded in me. I recognize it is somewhat necessary physically. But how do I set about realizing that psychologically it is not? How do I, who have had the habit of accumulating for millennia, general and particular, not only recognize the habit, but when I do recognize the habit, how does that movement come to an end? That is the real question.

Where does intelligence play a part in all this?

DB: There has to be intelligence to see this.

JK: Is it intelligence? Is it so-called ordinary intelligence or some other intelligence, something entirely different?

DB: I don't know what people mean by intelligence, but if they mean just merely the capacity to . . .

JK: To discern, to distinguish, to solve technical problems, economic problems, and so on. I would call that partial intelligence because it is not really . . .

DB: Yes, call that "skill in thought."

JK: All right, skill in thought. Now, wait a minute. That's what I'm trying to find out. I realize the reason for accumulation, division, security, the general and particular, thought. I can see the logic of all that. But logic, reason, and explanation don't end the thing. Another quality is necessary. Is that quality intelligence? I'm trying to move away from "insight" for a while. Is intelligence associated with thought? Is it related, is it part of thought, is it the outcome of very clear, precise, exact, logical conclusions of thought?

DB: That would still be more and more skill.

JK: Yes, skill.

DB: Yes, but when we say intelligence, at least we suggest the intelligence has a different quality.

JK: Yes. Is that intelligence related to love?

DB: I'd say they go together.

JK: Yes, I'm just moving slowly to that. You see, I've realized all that we have discussed, and I've come to a blank wall, a solid wall that I can't go beyond. And in observing, looking, fishing around,

I come upon this word "intelligence." And I see that the so-called intelligence of thought, skill, is not intelligence. So I'm asking further if this intelligence is associated with or related to or part of love. One cannot accumulate love.

DB: No, people might try. People do try to guarantee love.

JK: It sounds silly! That is all romantic nonsense, cinema stuff. You cannot accumulate love. You cannot associate it with hate. That love is something entirely different. And has that love intelligence? Which then operates? Which then breaks down the wall?

All right, sir, let's begin again. I don't know what that love is. I know all the physical bit. I realize pleasure, desire, accumulation, remembrance, images, are not love. I realized all that long ago. But I've come to the point where this wall is so enormous that I can't even jump over it. So I'm now fishing around to see if there is a different movement which is not a man-made movement. And that movement may be love. I'm sorry to use that word because it has been so spoilt and misused, but we'll use it for the time being.

So is that love, with its intelligence, the factor that will break down or dissolve or break up this wall? Not "I love you" or "you love me." It's not personal or particular. It's not general or particular. It is something beyond. I think when one loves with that intelligence, it covers the whole; it's not the particular or general. It is that. It is light; it's not particular light. If that is the factor that will break down the wall that is in front of me, then I don't know that love. As a human being, having reached a certain point, I can't go beyond it to find that love. What shall I do? Not "do" or "not do," but what is the state of my mind when I realize that any movement this side of the wall is still strengthening the wall? I realize, through meditation or whatever you do, that there is no movement, but the mind can't go beyond it.

But you come along and say, "Look, that wall can be dissolved, broken down, if you have that quality of love with intelligence." And I say, "Excellent, but I don't know what it is." What shall I do? I can't do anything; I realize that. Whatever I do is still on this side of the wall.

So am I in despair? Obviously not, because if I am in despair or depressed, I'm still moving in the same field. So all that has stopped. Realizing that I cannot possibly do anything, make any movement, what takes place in my mind? I realize I cannot do a thing. So what has happened to the quality of my mind, which has always moved to accumulate, to become? All that has stopped. The moment I realize this, no movement. Is that possible? Or am I living in illusion? Or have I really gone through all this to come to that point? Or do I suddenly say, I must be quiet?

Is there a revolution in my mind, a revolution in the sense that movement has completely stopped? And if it has, is love something beyond the wall?

DB: Well, it wouldn't mean anything.

JK: Of course, it couldn't be.

DB: The wall itself is the product of the process which is illusion.

JK: Exactly. I'm realizing that the wall *is* this movement. So when this movement ends, that quality of intelligence, love and so on, is there. That's the whole point.

DB: Yes. Could one say the movement ends, the movement sees that it has no point?

JK: It is like the so-called skill to see a danger.

DB: Well, it could be.

JK: Yes. Any danger demands a certain amount of awareness. But I have never realized as a human being that the accumulating process is a tremendous danger.

DB: Because that seems to be the essence of security.

JK: Of course. You come along and point it out to me, and I'm listening to you very carefully and I see, and I actually perceive the danger of that. And perception is part of love, isn't it? So the very perception, without any motive, without any direction, of the wall—which has been brought into being by this movement of accumulation—*is* intelligence and love.

Thirteen

The Ending of
"Psychological Knowledge"

18 SEPTEMBER 1980, BROCKWOOD PARK, HAMPSHIRE

JIDDU KRISHNAMURTI: What makes the mind always follow a certain pattern? Always seeking? If it lets go of one pattern, it picks up another; it keeps on functioning all the time like that. One can give explanations why it does so—for protection, for safety, from slackness, indifference, a certain amount of callousness, a total disregard of one's own flowering, etc.

It is really very important to find out why our minds are always operating in a certain direction.

We have said that one comes, after going through travail, investigation, and insight, to a blank wall. And that blank wall can only wither away or be broken down, when there is love and intelligence. But before we go into that, I would like to ask why human beings, however intelligent, however learned, however philosophical and religious, always fall into this groove.

DAVID BOHM: Well, I think the groove is inherent in the nature of the accumulated knowledge.

JK: Are you saying then that knowledge must invariably create a groove?

DB: Perhaps it is not inevitable, but it seems to develop this way in mankind, if we are referring to psychological knowledge, that is to say ...

JK: Obviously we are talking of that. But why does the mind not be aware of it, see the danger of this mechanical repetition and the fact that there is nothing new in it? See how we keep on doing it?

DB: It seems to me that the groove, or the accumulated knowledge, has a significance far beyond what its significance is. If we say that we have knowledge of some object, like the microphone, that has some limited significance. But knowledge about the nation to which you belong seems to have immense significance.

JK: Yes. So is this attribution of significance the cause of the narrowing down of the mind?

DB: Because this knowledge seems to have a tremendous value beyond all other values, it makes the mind stick to that. It seems the most important thing in the world.

JK: In India, there is this philosophy that knowledge must end—you know it, of course: the Vedanta. But apparently very, very few people do end knowledge and talk from freedom.

DB: You see, knowledge generally seems to be extremely important, even when a person may say verbally that it should end. . . . I mean knowledge about the self.

JK: You mean I am so stupid that I don't see that this psychological knowledge has very little significance, and so my mind clings to it?

DB: I wouldn't quite put it that a person is that stupid, but rather say that his knowledge stupefies the brain.

JK: Stupefies, all right. But the brain doesn't seem to extricate itself.

DB: It is already so stupefied that it can't see what it is doing.

JK: So what shall it do? I have been watching for many years people attempting to become free from certain things. This is the root of it. You understand? This psychological accumulation which becomes psychological knowledge. And so it divides, and all kinds of things happen around it and within it. And yet the mind refuses to let go.

DB: Yes.

JK: Why? Is that because there is safety or security in it?

DB: That is part of it, but I think in some way that knowledge has taken on the significance of the absolute, instead of being properly relative.

JK: I understand all that, but you are not answering my question. I am an ordinary man, I realize all this and the significance and the value of knowledge at different levels, but deeper down inside one, this accumulated knowledge is very, very destructive.

DB: The knowledge deceives the mind, so that the person is not normally aware that it is destructive. Once this process gets started, the mind is not in a state where it is able to look at it because it is avoiding the question. There is a tremendous defensive mechanism to escape from looking at the whole issue.

JK: Why?

DB: Because it seems that something extremely precious might be at stake.

JK: One is strangely intelligent, capable, or skilled in other directions, but here, where the root is of all this trouble, why don't we comprehend what is happening? What prevents the mind from doing this?

DB: Once importance has been given to knowledge, there is a mechanical process that resists intelligence.

JK: So what shall I do? I realize I must let go the accumulated, psychological knowledge—which is divisive, destructive, and rather petty—but I can't. Is this because of lack of energy?

DB: Not primarily, though the energy is being dissipated by the process.

JK: Having dissipated a great deal of energy, I haven't the energy to grapple with this?

DB: The energy would come back quickly if we could understand this. I don't think that is the main point.

JK: No. So what shall I do, realizing that this knowledge is inevitably forming a groove in which I live? How am I to break it down?

DB: Well, I am not sure that it is generally clear to people that this knowledge does all that, or that the knowledge *is* knowledge. You see, it may seem to be some "being," the "self," the "me." This knowledge creates the "me," and the "me" is experienced as an entity which seems not to be knowledge but some real being.

JK: Are you saying that this "being" is different from knowledge?

DB: It appears to be; it feigns a difference.

JK: But is it?

DB: It isn't, but the illusion has great power.

JK: That has been our conditioning

DB: Yes. Now, the question is how do we get through that to break down the groove, because it creates the imitation, or a pretension, of a state of being?

JK: That is the real point, you see. Look, there are millions of Catholics, a billion Chinese. This is their central movement. It seems so utterly hopeless. And realizing the hopelessness, I sit back and say I can't do anything. But if I apply my mind to it, the question arises, is it possible to function without psychological knowledge in this world? I am rather concerned about it; it seems the central issue that man must resolve, all over the world.

DB: That is right. You see, you may discuss this with somebody, who thinks it seems reasonable. But perhaps his status is threatened, and we have to say that that is psychological knowledge. It doesn't seem to him that it is knowledge, but something more. And he doesn't see that his knowledge of his status is behind the trouble. At first sight, knowledge seems to be something passive, which you could use if you wanted to and which you could just put aside if you wished, which is the way it should be.

JK: I understand all that.

DB: But then the moment comes when knowledge no longer appears to be knowledge.

JK: The politicians and the people in power wouldn't listen to this. And neither would the so-called religious people. It is only the people who are discontented, frustrated, who feel they have lost everything, who perhaps will listen. But they won't listen so that it is a real burning thing.

How does one go about this? Say, for instance, I have left Catholicism and Protestantism and all that. Also I have a career and I know that it is necessary to have knowledge there. Now, I see how important it is not to be caught in the process of psychological knowledge, and yet I can't let it go. It is always dodging me; I am playing tricks with it. It is like hide-and-seek. All right! We said that is the wall I have to break down. No, not I—that is the wall that has to be broken down. And we have said that this wall can be broken down through love and intelligence. Aren't we asking something enormously difficult?

DB: It is difficult.

JK: I am this side of the wall, and you are asking me to have that love and intelligence which will destroy it. But I don't know what that love is, what that intelligence is, because I am caught in this, on the other side of the wall. I realize logically, sanely, that what you are saying is accurate, true, logical, and I see the importance of it, but the wall is so strong and dominant and powerful that I can't get beyond it. We said the other day that the wall could be broken down through insight. You see, that insight becomes an idea.

DB: Yes.

JK: When insight is described to me, whether it is possible, how it is brought about, and so on, I make an abstraction of it, which means I move away from the fact, and the abstraction becomes all important. Which means, again, knowledge.

DB: Yes, the activity of knowledge.

JK: So we are back again!

DB: I think the general difficulty is that knowledge is not just sitting there as a form of information but is extremely active, meeting and shaping every moment according to past knowledge. So even when we raise this issue, knowledge is all the time waiting, and then acting. Our whole tradition is that knowledge is not active but passive. But it is really active, although people don't generally think of it that way. They think it is just sitting there.

JK: It is waiting.

DB: Waiting to act, you see. And whatever we try to do about it, knowledge is already acting. By the time we realize that this is the problem, it has already acted.

JK: Yes. But do I realize it as a problem, or as an idea which I must carry out? You see the difference?

DB: Knowledge automatically turns everything into an idea, which we must carry out. That is the whole way it is built.

JK: The whole way we have lived.

DB: Knowledge can't do anything else.

JK: How are we to break that, even for a second?

DB: It seems to me that if you could see, observe, be aware—if knowledge could be aware of itself at work . . . The point is that knowledge seems to work unawares, simply waiting, and then acting, by which time it has disrupted the order of the brain.

JK: I am very concerned about this, because wherever I go, this is what is happening. It is something that has to be resolved. Would you say the capacity to listen is far more important than any of this, than any explanations or logic?

DB: It comes to the same problem.

JK: No, no. It doesn't. I want to see if there is a possibility that when I listen completely to what you are saying, the wall has broken down. You understand? Is there—I am trying to find out—I am an ordinary man and you are telling me all this, and I realize what you are saying is so. I am really deeply involved in what you are saying, but somehow the flame isn't lit; all the fuel is there, but the fire is not. So, as an ordinary person, what shall I do? This is my everlasting cry!

DB: The brain has the capacity to listen; we have to question whether the ordinary man is so full of opinions that he can't listen.

JK: You can't listen with opinions; you might just as well be dead.

DB: I think knowledge has all sorts of defences. Is it possible for, say, the ordinary man to have this perception? That is really what you are asking, isn't it?

JK: Yes. But there must be a communication between you and that man, something so strong that the very act of his listening to you, and you communicating with him, operates.

DB: Yes, then you have to break through his opinions, through the whole structure.

JK: Of course. That is why this man has come here—for that. He has finished with all the churches and doctrines. He realizes that what has been said here is true, and he is burning to find out. When you communicate with him, your communication is strong and real, because you are not speaking from knowledge or opinions. A free human being is trying to communicate with this ordinary man. Now, can he listen with that intensity which you, the communicator, are giving him? He wants to listen to somebody who is telling the truth, and in the very telling of it, something is taking place in him. Because he is so ardently listening, this happens.

It is rather like you as a scientist, telling one of your students something. You are telling me about something which must be enormously important, because you have given your life to it. And I have given up so much just to come here. So is it the fault of you who are communicating with me that I do not receive it instantly? Or am I incapable of really listening to you?

DB: Well, if one is incapable of listening, what can be done? But let's say there is somebody who comes along who has got through some of these defences, although there are others that he is not aware of—that is something less simple than what you described.

JK: I feel it is dreadfully simple somehow. If one could listen with all one's being, the brain would not be caught in the groove. You see, generally, in communication, you are telling me something and I am absorbing it, but there is an interval between your telling and my absorbing.

DB: Yes.

JK: And that interval is the danger. If I didn't absorb but listened absolutely with all my being, it is finished. Now, is listening difficult because in this there is no shadow of pleasure? You are not offering any pleasure, any gratification. You are saying this is so; take it. But my mind is so involved in pleasure that it won't listen to anything that is not completely satisfactory or pleasurable.

I realize too the danger of that, of seeking satisfaction and pleasure, so I put that aside too. There is no pleasure, no reward, no punishment. In listening, there is only pure observation. So we come to the point: Is pure observation, which is actually listening, love? I think it is.

Again, if you state this, then my mind says, "Give it to me. Tell me what to do." But when I ask you to tell me what to do, I am back in the field of knowledge. It is so instantaneous. So I refuse to ask you what to do. Then where am I? You have told me perception without any motive or direction, pure perception, is love. And in that perception-love is intelligence. They are not three separate things; they are one thing. You pointed all this out very carefully, step by step, and I have come to the point that I have a feeling for it. But it goes away so quickly. Then the question begins, "How am I to get it back?" Again, the remembrance of it, which is knowledge, blocks.

DB: What you are saying is that every time there is a communication, knowledge gets to work in many different forms.

JK: So you see it is enormously difficult to be free of knowledge.

DB: We could ask, why doesn't knowledge wait until it is needed?

JK: That means to be psychologically free of knowledge, when the occasion arises, you act from freedom, not from knowledge.

DB: Knowledge comes in to inform your action, although it is not the source.

JK: That is freedom from knowledge. And being free, it is from freedom and not from knowledge that one communicates. That is, from emptiness there is communication. One may use words which are the outcome of knowledge, but it is from that state of complete freedom. Now, suppose I, as a human being, have come to that point where there is this freedom, and from it communication, using words, takes place. Will you, as an eminent scientist, communicate with me? Can I communicate with you, without any barrier? You follow what I am saying?

DB: Yes. There is this freedom from knowledge when knowledge is seen to be information. But ordinarily it seems more than information, and knowledge itself does not see that knowledge is not free.

JK: It is never free. And if I am going to understand myself, I must be free to look.

How will you communicate with me, who have come to a certain point where I am burning to receive what you are saying so completely that psychological knowledge is finished? Or am I fooling myself about being in that state?

DB: Well, that is the question. Knowledge is constantly deceiving itself.

JK: So is my mind always deceiving itself? Then what shall I do? Let's come back to that.

DB: Again, I think the answer is to listen.

JK: Why don't we listen? Why don't we immediately understand this thing? One can give all the superficial reasons why: old age, conditioning, laziness, and so on.

DB: But is it possible to give the deep reason for it?

JK: I think it is that the knowledge which is the "me" is so tremendously strong as an idea.

DB: Yes, that is why I tried to say that the idea has tremendous significance and meaning. For example, suppose you have the idea of God; this takes on a tremendous power.

JK: Or if I have the idea that I am British or French, it gives me great energy.

DB: And so it creates a state of the body which seems the very being of the self. And now the person doesn't experience it as mere knowledge . . .

JK: Yes, but are we going round and round and round? It seems like it.

DB: Well, I was wondering if there is anything that could be communicated about that overwhelming power that seems to come with knowledge . . .

JK: . . . and with identification.

DB: That seems to be something that would be worth looking into.

JK: Now, what is the root meaning of "identification"?

DB: Always the same.

JK: Always the same. That's right. That's right! There is nothing new under the sun.

DB: One says the self is always the same. It tries to be always the same in essence, if not in detail.

JK: Yes, yes.

DB: I think this is the thing that goes wrong with knowledge. It attempts to be knowledge of what is always the same, so it sticks, you see. Knowledge itself tries to find what is permanent and perfect. I mean, even independently of any of us. It is built into us, like the cells.

JK: From this arises the question, is it possible to attend diligently? I am using "diligence" in the sense of being accurate.

DB: Literally its root means to take pains.

JK: To take pains, of course. To take pain, take the whole of it. There must be some other way round all this intellectual business! We have exercised a great deal of intellectual capacity, and that has led to the blank wall. I approach it from every direction, but eventually the wall is there, which is the "me," with my knowledge, my prejudice, and all the rest of it. And the "me" then says, "I must do something about it." Which is still the "me."

DB: The "me" wants always to be the same, but at the same time it tries to be different.

JK: [*Laughs*] To put on a different coat. It is always the same. So the mind which is functioning with the "me" is always the same mind. Good Lord, you see, we are back again!

We have tried everything—fasting, every kind of thing—to get rid of the "me" with all its knowledge and illusions. One tries to identify with something else, which is the same. One then comes back to the fundamental question, what will make the wall totally disappear? I think this is only possible when I can give total attention to what the free man is saying. There is no other means to break down the wall—not the intellect, not the emotions, nor anything else. When somebody who has gone beyond the wall, who has broken it down, says, "Listen, for God's sake listen," and I listen to him with my mind empty, then it is finished. You know what I am saying? I have no sense of hoping for anything to happen or anything to come back or concern with the future. The mind is empty and therefore listening. It is finished.

To go on differently, for a scientist to discover something new he must have a certain emptiness from which there will be a different perception.

DB: Yes, but only in the sense that usually the question is limited, and so the mind may be empty with regard to that particular question, allowing the discovery of an insight in that area. But we are not questioning this particular area. We are questioning the whole of knowledge.

JK: It is most extraordinary when you go into it.

DB: And you were saying the end of knowledge is the Vedanta.

JK: That is the real answer.

DB: But generally people feel they must keep knowledge in one area to be able to question it in another. You see, it might worry people to ask, with what knowledge do I question the whole of knowledge?

JK: Yes. With what knowledge do I question my knowledge [*laughs*]? Quite.

DB: In a way, we do have knowledge, because we have seen that this whole structure of *psychological* knowledge makes no sense, that it is inconsistent and has no meaning.

JK: From that emptiness that we talked of previously, is there a ground or a source from which all things begin? Matter, human beings, their capacities, their idiocies—does the whole movement start from there?

DB: We could consider that. But let's try to clarify it a little. We have the emptiness.

JK: Yes, emptiness in which there is no movement of thought as psychological knowledge. And therefore no psychological time.

DB: Though we still have the time of the watch . . .

JK: Yes, but we have gone beyond that; don't let's go back to it. There is no psychological time, no movement of thought. And is that emptiness the beginning of all movement?

DB: Well, would you say the emptiness is the ground?

JK: That is what I am asking. Let's go slowly into this.

DB: Earlier on we were saying that there is the emptiness, and beyond that is the ground.

JK: I know, I know. Let's discuss this further.

Fourteen

The Mind in the Universe

JIDDU KRISHNAMURTI: We talked the other day about a mind that is entirely free from all movement, from all the things that thought has put there, the past, and the future, and so on. But before we go into that, I would like to discuss man's being caught in materialistic attitudes and values, and to ask, what is the nature of materialism?

DAVID BOHM: Well, first of all materialism is the name of a certain philosophical . . .

JK: I don't mean that.

DB: Matter is all there is, you see.

JK: I want to go into this. Nature, all human beings, react physically. This reaction is sustained by thought. And thought is a material process. So reaction in nature is the materialistic response.

DB: I think the word "materialistic" is not quite right. It is the response of matter.

JK: The response of matter—let's put it that way. That is better. We are talking about having an empty mind, and we have come to that point when the wall has been broken down. This emptiness and what lies beyond it, or through it—we will come to that, but before doing so, I am asking, is all reaction matter?

DB: Matter in movement. You could say that there is evidence in favour of that, that science has found a tremendous number of reactions which are due to the nerves.

JK: So would you say that matter and movement are the reactions which exist in all organic matter?

DB: Yes, all matter as we know it goes by the law of action and reaction, you see. Every action has a corresponding reaction.

JK: So action and reaction is a material process, as is thought. Now, to go beyond it is the issue.

DB: But before we say that, some people might feel that there is no meaning in going beyond it. That would be the philosophy of materialism.

JK: But if one is merely living in that area, it is very, very shallow. Right? It has really no meaning at all.

DB: Perhaps one should refer to one thing that some people have said, that matter is not merely action and reaction but may have a creative movement. You see, matter may create new forms.

JK: But it is still within that area.

DB: Yes. Let's try to make it clear. We have to see that there are very subtle forms of materialism which might be difficult to pin down.

JK: Let's begin. Would you consider that thought is a material process?

DB: Yes. Well, some people might argue that it is both material and something beyond material.

JK: I know. I have discussed this. But it is not.

DB: How can we say that, simply to make it clear?

JK: Because any movement of thought is a material process.

DB: Well, we have to make this clear so that it is not a matter of authority. As an observation, one sees that thought is a material process. Now, how are you going to see that?

JK: How could one be aware that thought is a material process? I think that is fairly clear. There is an experience, an incident which is recorded, which becomes knowledge. And from that knowledge, thought arises and action takes place.

DB: Yes. So we say that thought is that.

JK: Any assertion that is beyond is still thought.

DB: It is still coming from the background. So are you saying that something new coming about is not part of this process?

JK: Yes, if there is to be something new, thought, as a material process, must end. Obviously.

DB: And then it may take it up later.

JK: Later, yes. Wait. See what happens later. So could we say all action, reaction, and action from that reaction is movement of matter?

DB: Yes, very subtle movement of matter.

JK: So as long as one's mind is within that area, it must be a movement of matter. So is it possible for the mind to go beyond reaction? That, obviously, is the next step. As we said earlier, one gets irritated, and that is the first reaction. Then the reaction to that, the second reaction, is "I must not be." Then the third reaction is "I must control or justify." So it is constantly action and reaction. Can one see that this is a continuous movement without an ending?

DB: Yes. The reaction is continuous, but it seems at a certain moment to have ended, and the next moment appears to be a new movement.

JK: But it is still reaction.

DB: It is still the same, but it presents itself differently.

JK: It is exactly the same always . . .

DB: But it presents itself as always different, always new.

JK: Of course. That's just it. You say something, I get irritated, but that irritation is a reaction.

DB: Yes, it seems to be something suddenly new.

JK: But it is not.

DB: But one has to be aware of that, you see. Generally the mind tends not to be aware of it.

JK: But after discussing and talking about it, we are sensitive to it, alert to the question. So there is an ending to reaction if one is watchful, attentive; if one understands not only logically but has an insight into this constant reacting process, it can of course come to an end. That is why it is very important to understand this, before we discuss what is an empty mind, and if there is something beyond this, or whether in that very empty mind there is some other quality.

So is that empty mind a reaction? A reaction to the problems of pain and pleasure and suffering? An attempt to escape from all this into some state of nothingness?

DB: Yes, the mind can always do that.

JK: It can invent. That becomes an illusion. As we have said, desire is the beginning of illusion. Now, we have come to the point that this quality of emptiness is not a reaction. That must be absolutely sure, right? Now, before we go further, is it possible to have a mind that is really *completely empty* of all the things that thought has put together?

DB: Well, when thought ceases to react.

JK: That's it.

DB: On the one hand, perhaps you could say that reaction is due to the nature of matter, which is continually reacting and moving. But then is matter affected by this insight?

JK: I don't quite follow. Ah, I understand! Does that insight affect the cells of the brain which contain the memory?

DB: Yes. The memory is continually reacting, moving, as do air and water and everything around us.

JK: After all, if I don't react physically, I am paralysed. But to be reacting continuously is also a form of paralysis.

DB: Well, the wrong kind of reaction! Reaction around the psychological structure. But assuming that the reaction around the psychological structure has begun in mankind, why should it ever stop? Because reaction makes another and another, and one would expect it to go on forever unless something will stop it.

JK: Nothing will stop it. Only insight into the nature of reaction ends psychological reaction.

DB: Then you are saying that matter is affected by insight, which is beyond matter.

JK: Yes, beyond matter. So is this emptiness within the brain itself? Or is it something that thought has conceived as being empty? One must be very clear.

DB: Yes. But whatever we discuss, no matter what the question is, thought begins to want to do something about it, because thought feels it can always make a contribution.

JK: Quite.

DB: So thought in the past did not understand that there are areas where it has no useful contribution to make, but it has kept on in the habit of trying to say that emptiness is very good. Therefore thought says, I will try to bring about the emptiness.

JK: Of course.

DB: Thought is trying to be helpful!

JK: We have been through all that. We have seen the nature of thought, and its movement, time, and all that. But now I want to find out whether this emptiness is within the mind itself or beyond it.

DB: What do you mean by the mind?

JK: The mind is the whole—emotions, thought, consciousness, the brain—the whole of that is the mind.

DB: The word "mind" has been used in many ways. Now you are using it in a certain way, that it represents thought, feeling, desire, and will—the whole material process.

JK: Yes, the whole material process.

DB: Which people have called non-material!

JK: Quite. But the mind is the whole material process.

DB: Which is going on in the brain and the nerves.

JK: The whole structure. One can see that this materialistic reaction can end. And the next question I am asking is whether that emptiness is within or without. "Without" in the sense of being elsewhere.

DB: Where would it be?

JK: I don't think it would be elsewhere, but I am just putting the question . . .

DB: Well, any such thing is part of the material process. "Here" and "there" are distinctions made in the material process.

JK: Yes, that is right. It is in the mind itself. Not outside it. Right?

DB: Yes.

JK: Now, what is the next step? Does that emptiness contain nothing? Not a thing?

DB: Not a thing, by which we mean anything that has form, structure, stability.

JK: Yes. All that—form, structure, reaction, stability, capacity. None of that. Then what is it? Is it then total energy?

DB: Yes, movement of energy.

JK: Movement of energy. It is not movement of reaction.

DB: It is not movement of things reacting to each other. The world can be regarded as made up of a large number of things which react to each other, and that is one kind of movement, but we are saying it is a different kind of movement.

JK: Entirely different.

DB: Which has no thing in it.

JK: No thing in it, and therefore it is not of time. Is that possible? Or are we just indulging in imagination? In some kind of romantic, hopeful, pleasurable sensation? I don't think that we are, because we have been through all that, step by step, right up to this point. So we are not deceiving ourselves. Now, we say that emptiness has no centre as the "me" and all the reactions. In that emptiness there is a movement of timeless energy.

DB: When you refer to timeless energy, we could repeat what we have already said about time and thought being the same.

JK: Yes, of course.

DB: Then you were saying that time can only come into a material process.

JK: That's right.

DB: Now, if we have energy, that is timeless but nevertheless moving . . .

JK: Yes, not static . . .

DB: Then what is the movement?

JK: What is movement? From here to there.

DB: That is one form.

JK: One form. Or from yesterday to today and from today to tomorrow.

DB: Yes, there are various kinds of movement.

JK: So what is movement? Is there a movement which is not moving? You understand? Is there a movement which has no beginning and no end? Because thought has a beginning and an end.

DB: Except you could say that the movement of matter might have no beginning and no ending—the reactive movement. You are not speaking of that?

JK: No, I am not talking of that. Thought has a beginning and thought has an ending. There is a movement of matter as reaction and the ending of that reaction.

DB: In the brain.

JK: Yes, there are these various kinds of movement. That is all we know. And someone comes along and says there is a totally different kind of movement. But to understand that, I must be free of the movement of thought and the movement of time, to understand a movement that is not . . .

DB: Well, there are two things about this movement. It has no beginning and no end, but also it is not determined as a series of successions from the past.

JK: Of course. No causation.

DB: But, you see, matter can be looked at as a series of causes; it may not be adequate. But now you are saying that this movement has no beginning and no end; it is not the result of a series of causes following one another.

JK: So I want to understand, verbally even, a movement that is not a movement. I don't know if I am making it clear.

DB: Then why is it called a movement if it is not a movement?

JK: Because it is not still; it is active, dynamic.

DB: It is energy.

JK: It is tremendous; therefore it can never be still. But it has got in that energy a stillness.

DB: I think we have to say that ordinary language does not convey this properly, but the energy itself is still and also moving.

JK: But in that movement it is a movement of stillness. Does it sound crazy?

DB: The movement can be said to emerge from stillness and draw back into stillness.

JK: That's right. You see, that is what it is. We said that this emptiness is in the mind. It has no cause and no effect. It is not a movement of thought, of time. It is not a movement of material reactions. None of that. Which means is the mind capable of that extraordinary stillness without any movement? And when it is so completely still, there is a movement out of it. It sounds crazy!

DB: Well, it needn't sound crazy. I think I mentioned before that some people, like Aristotle, had this notion in the past; we discussed it. He talked about the unmoved mover, when trying to describe God, you see.

JK: Ah, God, no. I am not describing . . .

DB: You don't want to describe God, but some sort of notion similar to this has been held in the past by various people. Since then it has gone out of fashion, I think.

JK: Let's bring it into fashion, shall we?!

DB: I am not saying that Aristotle had the right idea. It is merely that he was considering something somewhat similar, though probably different in many respects.

JK: Was it an intellectual concept or an actuality?

DB: This is very hard to tell because so little is known.

JK: Therefore we don't have to bring in Aristotle.

DB: I merely wanted to point out that the concept of a movement out of stillness wasn't crazy, because other very respectable people had had something similar.

JK: I am glad! I am glad to be assured that I am not crazy [*laughs*]! And is that movement out of stillness the movement of creation? We are not talking of what the poets, writers, and painters call creation. To me, that is not creation, just capacity, skill, memory, and knowledge operating. Here I think this creation is not expressed in form.

DB: It is important to differentiate. Usually we think creation is expressed in form, or as structure.

JK: Yes, structure. We have gone beyond being crazy, so we can go on! Would you say that this movement, not being of time, is eternally new?

DB: Yes. It is eternally new in the sense that the creation is eternally new. Right?

JK: Creation is eternally new. You see, that is what the artists are trying to find. Therefore they indulge in all kinds of absurdities, but to come to that point where the mind is absolutely silent, and out of that silence there is this movement which is always new. . . . And the moment when that movement is expressed . . .

DB: . . . the first expression is in thought.

JK: That is just it.

DB: And that may be useful, but then it gets fixed and may become a barrier.

JK: I was told once by an Indian scholar that before people began to sculpt the head of a god, or whatever, they had to go into deep meditation. At the right moment they took up the hammer and chisel.

DB: To have it come out of the emptiness. There is another point, you see. The Australian aborigines draw figures in the sand, so that they don't have permanency.

JK: That is right.

DB: Perhaps thought could be looked at that way. You see, marble is too static and remains for thousands of years. So although the original sculptor may have understood, the people who follow see it as a fixed form.

JK: So what relationship has all this to my daily life? In what way does it act through my actions, through my ordinary physical responses, to noise, to pain, various forms of disturbance? What relationship has the physical to that silent movement?

DB: Well, insofar as the mind is silent, the thought is orderly.

JK: We are getting onto something. Would you say that silent movement, with its unending newness, is total order of the universe?

DB: We could consider that the order of the universe emerges from this silence and emptiness and is eternally creative.

JK: So what is the relationship of this mind to the universe?

DB: The particular mind?

JK: No, mind.

DB: Mind in general?

JK: Mind. We went through the general and the particular, and beyond that there is the mind.

DB: Would you say that is universal?

JK: I don't like to use the word "universal."

DB: Universal in the sense of that which is beyond the particular and general. But perhaps that word is difficult.

JK: Can we find another word? Not global. A mind that is beyond the particular?

DB: Well, you could say it is the source, the essence. It has been called the absolute.

JK: I don't want to use the word "absolute" either.

DB: The absolute means literally that which is free of all limitations, of all dependence.

JK: All right, if you agree that "absolute" means freedom from all dependence and limitation.

DB: From all relationships.

JK: Then we will use that word.

DB: It has unfortunate connotations.

JK: Of course. But let's use it for the moment just for convenience in our dialogue. There is this absolute stillness, and in or from that stillness there is a movement, and that movement is everlastingly new. What is the relationship of that mind to the universe?

DB: To the universe of matter?

JK: To the whole universe: matter, trees, nature, man, the heavens.

DB: That is an interesting question.

JK: The universe is in order; whether it is destructive or constructive, it is still order.

DB: You see, the order has the character of being absolutely necessary; in a sense it cannot be otherwise. The order that we usually know is not absolutely necessary. It could be changed; it could depend on something else.

JK: The eruption of a volcano is order.

DB: It is order of the whole universe.

JK: Quite. Now, in the universe there is order, and this mind which is still is completely in order.

DB: The deep mind, the absolute.

JK: The absolute mind. So is this mind the universe?

DB: In what sense is that the universe? We have to understand what it means to say that, you see.

JK: It means is there a division, or a barrier, between this absolute mind and the universe? Or are both the same?

DB: Both are the same.

JK: That is what I want to get at.

DB: We have either duality of mind and matter, or they are both the same.

JK: That's it. Is that presumptuous?

DB: Not necessarily. I mean that these are just two possibilities.

JK: I want to be quite sure that we are not treading upon something which really needs a very subtle approach, which needs great care. You know what I mean?

DB: Yes. Let's go back to the body. We have said that the mind, which is in the body—thought, feeling, desire, the general, and the particular mind—is part of the material process.

JK: Absolutely.

DB: And not different from the body.

JK: That's right. All the reactions are material processes.

DB: And therefore what we usually call the mind is not different from what we usually call the body.

JK: Quite.

DB: Now, you are making this much greater in saying, consider the whole universe. And we ask if what we call the mind in the universe is different from what we call the universe itself and matter?

JK: That's right. You see, that's why I feel that in our daily life there must be order, but not the order of thought.

DB: Well, thought is a limited order; it is dependent, relative.

JK: That's it. So there must be an order that is . . .

DB: . . . free of limitation.

JK: Yes. In our daily life we have to have that, which means no conflict, no contradiction whatsoever.

DB: Let's take the order of thought. When it is rational, it is in order. But in contradiction the order of thought has broken down, it has reached its limit. Thought works until it reaches a contradiction, and that's the limit.

JK: So if in my daily life there is complete order, in which there is no disturbance, what is the relationship of that order to the never-ending order? Can that silent movement of order, of that extraordinary something, affect my daily life, when I have deep inward psychological order? You understand my question?

DB: Yes. We have said, for example, that the volcano is a manifestation of the whole order of the universe.

JK: Absolutely. Or a tiger killing a deer.

DB: The question then is whether a human being in his daily life can be similar.

JK: That's it. If not, I don't see what is the point of the other.

DB: Well, it has no point to the human being. Then he falls back into trying to make his own purpose out of himself, out of his thoughts. You see, some people would say, "Who cares about the universe? All we care about is our own society and what we are doing." But then that falls down, because it is full of contradiction.

JK: Obviously. It is only thought which says that. So that universe, which is in total order, *does* affect my daily life.

DB: Yes. I think that scientists might ask how. You see, one might say, I understand that the universe is constituted of matter and that the laws of matter affect our daily life. But it is not so clear how it affects the mind, and that there is this absolute mind which affects the daily life.

JK: Ah! What is my daily life? A series of reactions and disorders, right?

DB: Well, it is mostly that.

JK: And thought is always struggling to bring order within that. But when it does that, it is still disorder.

DB: Because thought is always limited by its own contradictions.

JK: Of course. Thought is always creating disorder, because it is itself limited.

DB: As soon as it tries to go beyond the limit, that is disorder.

JK: Right. I have understood, I have gone into it, I have an insight into it, so I have a certain kind of order in my life. But that order is still limited. I recognize that, and I say that this existence is limited.

DB: Now, some people would accept that and say, "Why should you have more?"

JK: I am not "having more."

DB: They would say, "We would be happy if we could live in the ordinary material life with real order."

JK: I say let's do it! Of course, that must be done. But in the very doing of it, one has to realize it is limited.

DB: Yes, even the highest order we can produce is limited.

JK: And the mind realizes its limitation and says, let's go beyond it.

DB: Why? Some people would say, why not be happy within those limits, continually extending them, trying to discover new thoughts, new order? The artist will discover new forms of art, the scientists a new kind of science.

JK: But all that is always limited.

DB: Some people would go this far and say that this is all that is possible.

JK: I like, accept, and make the most of the human condition?

DB: Well, people would say that man could do much better than he is doing.

JK: Yes, but all this is still the human condition, a little reformed, a little better.

DB: Some people would say enormously reformed.

JK: But it is still limited!

DB: Yes. Let's try to make clear what is wrong with the limitation.

JK: In that limitation there is no freedom, only a limited freedom.

DB: Yes. So eventually we come to the boundary of our freedom. Something unknown to us makes us react, and through reaction we fall back into contradiction.

JK: Yes, but what happens when I see that I am always moving within a certain area?

DB: Then I am under the control of the forces.

JK: The mind inevitably rebels against that.

DB: That is an important point. You say the mind wants freedom. Right?

JK: Obviously.

DB: It says that freedom is the highest value. So do we accept that and see it just as a fact?

JK: That is, I realize that within this limitation I am a prisoner.

DB: Some people get used to it and say, "I accept it."

JK: I won't accept it! My mind says there must be freedom from prison. I am a prisoner, and the prison is very nice, very cultured, and all the rest of it. But it is still limited, and it says there must be freedom beyond all this.

DB: Which mind says this? Is it the particular mind of the human being?

JK: Ah! Who says there must be freedom? Oh, that is very simple. The very pain, the very suffering demands that we go beyond.

DB: This particular mind, even though it accepts limitation, finds it painful.

JK: Of course.

DB: And therefore this particular mind feels somehow that it is not right. But it can't avoid it. There seems to be a necessity of freedom.

JK: Freedom is necessary, and any hindrance to freedom is retrogression. Right?

DB: That necessity is not an external necessity due to reaction.

JK: Freedom is not a reaction.

DB: The necessity of freedom is not a reaction. Some people would say that having been in prison you reacted in this way.

JK: So where are we? You see, this means there must be freedom from reaction, freedom from the limitation of thought, freedom from all the movement of time. We know that there must be complete freedom from all that before we can really understand the empty mind and the order of the universe, which is then the order of the mind. You are asking of me a tremendous lot. Am I willing to go that far?

DB: Well, you know that non-freedom has its attractions.

JK: Of course, but I am not interested in these attractions.

DB: But you asked the question: Am I willing to go that far? So it seems to suggest that there may be something attractive in this limitation.

JK: Yes, I have found safety, security, pleasure in non-freedom. I realize that in that pleasure, pain, there is no freedom. The mind says, not as a reaction, that there must be freedom from all this. To come to that point and to let go without conflict demands its own discipline, its own insight. That's why I have said to those of us who have done a certain amount of investigation into all this, can one go as far as that? Or do the responses of the body—the responsibilities of daily action, for one's wife, children, and all that— prevent this sense of complete freedom? The monks, the saints, and the sannyasis have said, "You must abandon the world."

DB: We went into that. They take the world with them anyway.

JK: Yes. That is another form of idiocy, although I'm sorry to put it like that. We have been through all that, so I refuse to enter again into it. Now I say, are the universe and the mind that has emptied itself of all this, are they one?

DB: Are they one?

JK: They are not separate; they are one.

DB: It sounds as if you are saying that the material universe is like the body of the absolute mind.

JK: Yes, all right, all right.

DB: It may be a picturesque way of putting it!

JK: We must be very careful also not to fall into the trap of saying that the universal mind is always there.

DB: How would you put it then?

JK: They have said that God is always there; Brahman, or the highest principle, is always present, and all you have to do is to cleanse yourself and arrive at that. This is also a very dangerous statement, because then you might say there is the eternal in me.

DB: Well, I think thought is projecting.

JK: Of course!

DB: There is a logical difficulty in saying it is always there because "always" implies time, and we are trying to discuss something that has nothing to do with time. So we can't place it as being here, there, now, or then!

JK: So we have come to a point that there is this universal mind, and the human mind can be of that when there is freedom.

Fifteen

Can Human Problems
Be Solved?

27 SEPTEMBER 1980, BROCKWOOD PARK, HAMPSHIRE

JIDDU KRISHNAMURTI: We have cultivated a mind that can solve almost any technological problem. But apparently human problems have never been solved. Human beings are drowned by their problems: the problems of communication, knowledge, of relationships, the problems of heaven and hell. The whole of human existence has become a vast, complex problem. And apparently throughout history it has been like this. In spite of his knowledge, in spite of his centuries of evolution, man has never been free of problems.

DAVID BOHM: Yes, I would add, of insoluble problems.

JK: I question if human problems are insoluble.

DB: I mean as they are put now.

JK: As they are now, of course, these problems have become incredibly complex and insoluble. No politician, scientist, or philosopher is going to solve them, even through wars and so on! So why has the mind of human beings throughout the world not been able to resolve the daily problems of life? What are the things that prevent the complete solution of these problems? Is it that we have never turned our minds to it? Because we spend all our days, and probably half the night, in thinking about technological problems so that we have no time for the other?

DB: That is partly so. Many people feel that the other should take care of itself.

JK: But why? I am asking in this dialogue whether it is possible to have no human problems at all—only technological problems, which can be solved. But human problems seem insoluble. Why? Is it because of our education, our deep-rooted tradition, that we accept things as they are?

DB: Well, that is certainly part of it. These problems accumulate as civilization gets older, and people keep on accepting things which make problems. For example, there are now far more nations in the world than there used to be, and each one creates new problems.

JK: Of course.

DB: If you go back in time . . .

JK: . . . a tribe becomes a nation.

DB: And then the group must fight its neighbour.

JK: And they have this marvellous technology to kill each other. But we are talking about human problems of relationship, problems of lack of freedom, this sense of constant uncertainty and fear, the human struggle to work for a livelihood for the rest of one's life. The whole thing seems so extraordinarily wrong.

DB: I think people have lost sight of that. Generally speaking they accept the situation in which they find themselves and try to make the best of it, trying to solve some small problems to ameliorate their circumstances. They wouldn't even look at this whole situation seriously.

JK: And the religious people have created a tremendous problem for man.

DB: Yes. They are trying to solve problems too. I mean everybody is caught up in his own little fragment, solving whatever he thinks he can solve, but it all adds up to chaos.

JK: To chaos and wars! That is what we are saying. We live in chaos. But I want to find out if I can live without a single problem for the rest of my life. Is that possible?

DB: Well, I wonder if we should even call these things problems, you see. A problem would be something that is reasonably solvable. If you put the problem of how to achieve a certain result, then that presupposes that you can reasonably find a way to do it technologically. But psychologically the problem cannot be looked at in that way; to propose a result, you have to achieve and then find a way to do it.

JK: What is the root of all this? What is the cause of all this human chaos? I am trying to come to it from a different angle, to discover whether there is an ending to problems. You see, personally, I refuse to have problems.

DB: Somebody might argue with you about that and say that maybe you are not challenged with something.

JK: I was challenged the other day about something very, very serious.

DB: Then it is a matter of clarification. Part of the difficulty is clarification of the language.

JK: Clarification not only of language but of relationship and action. A problem arose the other day which involved lots of people, and a certain action had to be taken. But to me personally it was not a problem.

DB: We have to make it clear what you mean, because without an example, I don't know.

JK: I mean by a problem something that has to be resolved, something you worry about; something you are questioning and endlessly concerned with. Also doubts and uncertainties, and having to take some kind of action which you will regret at the end.

DB: Let's begin with the technical problem where the idea first arose. You have a challenge, something which needs to be done, and you say that is a problem.

JK: Yes, that is generally called a problem.

DB: Now, the word "problem" is based on the idea of putting forth something—a possible solution—and then trying to achieve it.

JK: Or I have a problem but I don't know how to deal with it.

DB: If you have a problem and you have no idea how to deal with it . . .

JK: . . . then I go round asking people for advice, and getting more and more confused.

DB: This would already be a change from the simple idea of a technical problem, where you usually have some notion of what to do.

JK: Surely technical problems are fairly simple.

DB: They often bring challenges requiring us to go very deeply and change our ideas. With a technical problem, we generally know what we have to do to solve it. For example, if there is lack of food, what we have to do is to find ways and means of producing more. But with a psychological problem, can we do the same?

JK: That is the point. How do we deal with this thing?

DB: Well, what kind of problem shall we discuss?

JK: Any problem which arises in human relationships.

DB: Let's say that people cannot agree; they fight each other constantly.

JK: Yes, let's take that for a simple thing. It seems to be almost impossible for a group of people to think together, to have the same outlook and attitude. I don't mean copying each other, of course. But each person puts his opinion forward and is contradicted by another, which goes on all the time, everywhere.

DB: All right. So can we say that our problem is to work together, to think together?

JK: Work together, think together, cooperate without monetary issues being involved in it.

DB: That is another question, that people will work together if they are highly paid.

JK: So how do we solve this problem? In a group, all of us are offering different opinions, and so we don't meet each other at all. And it seems almost impossible to give up one's opinions.

DB: Yes, that is one of the difficulties, but I am not sure that you can regard it as a problem and ask what shall I do to give up my opinions!

JK: No, of course. But that is a fact. So observing that and seeing the necessity that we should all come together, people still cannot give up their opinions, their ideas, their own experiences and conclusions.

DB: Often it may not seem to them like an opinion but the truth.

JK: Yes, they would call it fact. But what can man do about these divisions? We see the necessity of working together—not for some ideal, belief, some principle or some god. In countries throughout the world and even in the United Nations, they are not working together.

DB: Some people might say that we not only have opinions but self-interest. If two people have conflicting self-interests, there is no way, as long as they maintain their attachment to them, that they can work together. So how do we break into this?

JK: If you point out to me that we must work together, and show me the importance of it, I can also see that it is important. But I can't do it!

DB: That's the point. It is not enough even to see that cooperation is important and to have the intention of achieving this. With this inability there is a new factor coming in. Why is it that we cannot carry out our intentions?

JK: One can give many reasons for that, but those causes and reasons and explanations don't solve the problem. We come back to the same thing: What will make a human mind change? We see that change is necessary and yet are incapable or unwilling to change. What factor—what new factor—is necessary for this?

DB: Well, I feel it is the ability to observe deeply whatever it is that is holding the person and preventing him from changing.

JK: So is the new factor attention?

DB: Yes, that is what I meant. But also we have to consider what kind of attention.

JK: First let's discuss what attention is.

DB: It may have many meanings to different people.

JK: Of course, as usual, there are so many opinions! Where there is attention, there is no problem. Where there is inattention, every difficulty arises. Now, without making attention itself into a problem, what do we mean by it? So that I understand not verbally, not intellectually, but deeply, in my blood, the nature of attention in which no problem can ever exist. Obviously attention is not concentration. It is not an endeavour, an experience, a struggle to be attentive. But you show me the nature of attention, which is that when there is attention, there is no centre from which "I" attend.

DB: Yes, but that is the difficult thing.

JK: Don't let's make a problem of it!

DB: I meant that I have been trying this for a long time. I think that there is first of all some difficulty in understanding what is meant by attention, because of the content of thought itself. When a person is looking at it, he may think he is attending.

JK: No, in that state of attention there is no thought.

DB: But how do you stop thought then? You see, while thinking is going on, there is an impression of attention, which is not attention. But one thinks, one supposes that one is paying attention.

JK: When one supposes one is paying attention, that is not it.

DB: So how do we communicate the true meaning of attention?

JK: Or would you say rather that to find out what is attention, we should discuss what is inattention?

DB: Yes.

JK: And through negation come to the positive. When I am inattentive, what takes place? In my inattentiveness, I feel lonely, desperate, depressed, anxious, and so on.

DB: The mind begins to break up and go into confusion.

JK: Fragmentation takes place. Or in my lack of attention, I identify myself with many other things.

DB: Yes, and it may also be pleasant, but it can be painful too.

JK: I find, later on, that what was pleasing becomes pain. So all that is a movement in which there is no attention. Right? Are we getting anywhere?

DB: I don't know.

JK: I feel that attention is the real solution to all this—a mind which is really attentive, which has understood the nature of inattention and moves away from it!

DB: But first, what is the nature of inattention?

JK: Indolence, negligence, self-concern, self-contradiction—all that is the nature of inattention.

DB: Yes. You see, a person who has self-concern may feel that he is attending, but he is simply concerned with himself.

JK: Yes. If there is self-contradiction in me, and I pay attention to it in order not to be self-contradictory, that is not attention.

DB: But can we make this clear, because ordinarily one might think that this is attention.

JK: No, it is not. It is merely a process of thought, which says, "I am this; I must be that."

DB: So you are saying that this attempt to become is not attention.

JK: Yes, that's right. Because the psychological becoming breeds inattention.

DB: Yes.

JK: Isn't it very difficult, sir, to be free of becoming? That is the root of it. To end becoming.

DB: Yes. There is no attention, and that is why these problems are there.

JK: Yes, and when you point that out, the paying attention also becomes a problem.

DB: The difficulty is that the mind plays tricks, and in trying to deal with this, it does the very same thing again.

JK: Of course. That mind, which is so full of knowledge, self-importance, self-contradiction, and all the rest of it, has come to a point where it finds itself psychologically unable to move.

DB: There is nowhere for it to move.

JK: What would I say to a person who has come to that point? I come to you. I am full of this confusion, anxiety, and sense of despair, not only for myself but for the world. I come to that point, and I want to break through it. So it becomes a problem to me.

DB: Then we are back; there is again an attempt to become, you see.

JK: Yes. That is what I want to get at. So is that the root of all this? The desire to become?

DB: Well, it must be close to it.

JK: So how do I look, without the movement of becoming, at this whole complex issue of myself?

DB: It seems that one hasn't looked at the whole. We did not look at the whole of becoming, when you said, "How can I pay attention?" Part of it seemed to slip out and became the observer. Right?

JK: Psychologically, becoming has been the curse of all this. A poor man wants to be rich, and a rich man wants to be richer; it is all the time this movement of becoming, both outwardly and inwardly. And though it brings a great deal of pain and sometimes pleasure, this sense of becoming, fulfilling, achieving psychologically, has made my life into all that it is. Now, I realize that, but I can't stop it.

DB: Why can't I stop it?

JK: Let's go into that. Partly I am concerned in becoming because there is a reward at the end of it; also I am avoiding pain or punishment. And in that cycle I am caught. That is probably one of the reasons why the mind keeps on trying to become something. And the other, perhaps, is deep-rooted anxiety or fear that if I don't become, be something, I am lost. I am uncertain and insecure, so the mind has accepted these illusions and says I cannot end that.

DB: But why doesn't the mind end it? Also we have to go into the question of there being no meaning to these illusions.

JK: How do you convince me that I am caught in an illusion? You can't, unless I see it myself. I cannot see it because my illusion is so strong. That illusion has been nurtured, cultivated by religion, by the family, and so on. It is so deeply rooted that I refuse to let it go. That is what is taking place with a large number of people. They say, "I want to do this but I cannot." Now, given that situation, what is one to do? Will explanations, logic, and all the various logical contradictions help them? Obviously not.

DB: Because it all gets absorbed into the structure.

JK: So what is the next thing?

DB: You see, if they say, "I want to change," there is also the wish not to change.

JK: Of course. The man who says "I want to change" has also at the back of his mind, "Really, why should I change?" They go together.

DB: So we have a contradiction.

JK: I have lived in this contradiction, I have accepted it.

DB: But why should I have accepted it?

JK: Because it is a habit.

DB: But when the mind is healthy, it will not accept a contradiction.

JK: But our mind isn't healthy! The mind is so diseased, so corrupt, so confused that even though you point out all the dangers of this, it refuses to see them.

So how do we help a man who is caught in this to see clearly the danger of psychological becoming? Let's put it that way. Psychologically, becoming implies identification with a nation, a group, and all that business.

DB: Yes, holding to opinions.

JK: Opinions and beliefs. I have had an experience, it gives me tremendous satisfaction, I am going to hold on to it. How do you help me to be free of all this? I hear your words, your explanations, logic; they seem quite right, but I can't move out of all that.

I wonder if there is another factor, another way of communication, which isn't based on words, knowledge, explanations, and reward and punishment. Is there another way of communicating? You see, in that too there is a danger. Yet I am sure there is a way which is not verbal, analytical, or logical, which doesn't mean lack of sanity.

DB: Perhaps there is.

JK: My mind has always communicated with another with words, explanations and logic, or with suggestion. There must be another element which breaks through all that.

DB: It will break through the inability to listen.

JK: Yes, the inability to listen, the inability to observe, to hear, and so on. There must be a different method. You see, I have met several men who have been to a certain saint, and in his company they say all their problems are resolved. But when they go back to their daily life, they are back in the old game.

DB: There was no intelligence in it, you see.

JK: That is the danger. That man, that saint being quiet and non-verbal, in his very presence they feel quiet and feel their problems are resolved.

DB: But this is still from the outside.

JK: Of course. It is like going to church. In an ancient church or cathedral, you feel extraordinarily quiet. It is the atmosphere, the structure—you know. The very atmosphere makes you be quiet.

DB: Yes, it communicates what is meant by quietness, nonverbally.

JK: That is nothing. It is like incense!

DB: It is superficial.

JK: Utterly superficial: Like incense, it evaporates! So we push all that aside, and then what have we left? Not an outside agency, a god, or some saviour. What have I left? What is there that can be communicated, which will break through the wall that human beings have built for themselves?

Is it love? That word has become corrupted, loaded, dirty. But cleansing that word, is love the factor that will break through all this clever analytical approach? Is love the element that is lacking?

DB: Well, we have to discuss it; perhaps people are somewhat chary of that word.

JK: I am chary beyond words!

DB: And therefore as people resist listening, they will resist love too.

JK: That is why I said it is rather a risky word.

DB: We were saying the other day that love contains intelligence.

JK: Of course.

DB: Which is care as well; we mean by love that energy which also contains intelligence and care—all that . . .

JK: Now, wait a minute. You have that quality, and I am caught in my misery, anxiety, and so on. And you are trying to penetrate with that intelligence this mass of darkness. How will you do it? Will that act? If not, we human beings are lost. You follow, sir? Therefore we have invented Jesus, Buddha, Krishna, who love you—images which have become meaningless, superficial, and nonsensical.

So what shall I do? I think that is the other factor. Attention, perception, intelligence, and love—you bring all this to me, and I am incapable of receiving it. I say, "It sounds nice; I feel it, but I can't hold it." I can't hold it, because the moment I go outside this room, I am lost!

DB: That really is the problem.

JK: Yes, that is the real problem. Is love something outside, as a saviour, heaven—all that stuff outside? Is love something outside of me, which you bring to me, which you awaken in me, which you give me as a gift? Or, in my darkness, illusion, and suffering, is there that quality? Obviously not. There can't be.

DB: Then where is it?

JK: That's just it. Love is not yours or mine; it is not personal, not something that belongs to one person and not to the other. Love is not that.

DB: That is an important point. Similarly you were saying that isolation does not belong to any one person, although we tend to think of isolation as my personal problem.

JK: Of course. It is common ground for all of us. Also intelligence is not personal.

DB: But, again, that goes contrary to the whole of our thinking, you see.

JK: I know.

DB: Everybody says this person is intelligent and that one is not. And saying if I have a lack of intelligence, I must acquire it for myself. So this may be one of the barriers to the whole thing, that behind the ordinary everyday thought there is a deeper thought of mankind, which is that we are all divided and that these various qualities either belong to us or don't belong to us.

JK: Quite. It is the fragmented mind that invents all that.

DB: It has been invented, but we have picked it up verbally and nonverbally, by implication, from childhood. Therefore it pervades; it is the ground of our thoughts, of all our perceptions. So this has to be questioned.

JK: We have questioned it—that grief is not my grief, grief is human, and so on.

DB: But how are people to see that? Because a person who is experiencing grief feels that it is his personal grief.

JK: I think it is partly because of our education, partly our society and traditions.

DB: But it is also implicit in our whole way of thinking. Then we have to jump out of that, you see.

JK: Yes. But to jump out of that becomes a problem, and then what am I to do?

DB: Perhaps we can see that love is not personal.

JK: Earth is not English earth or French earth. Earth is earth!

DB: I was thinking of an example in physics: If a scientist or chemist is studying an element such as sodium, it's not that he's studying his sodium and somebody else is studying *his* and they somehow compare notes.

JK: Quite. Sodium is sodium.

DB: Sodium is sodium universally. So we have to say that love is love universally.

JK: Yes, but, you see, my mind refuses to see that, because I am so terribly personal, terribly "me and my problems" and all that. I refuse to let that go. When you say sodium is sodium, it is very simple; I can see that. But when you say to me that grief is common to all of us ... Sodium is grief [*laughs*]!

DB: This can't be done with time. But, you see, it took quite a while for mankind to realize that sodium is sodium.

JK: Is love something that is common to all of us?

DB: Well, insofar as it exists, it has to be common.

JK: Of course.

DB: It may not exist, but if it does, it has to be common.

JK: I am not sure it does not exist. Compassion is not "I am compassionate." Compassion is there, is something that is not "me compassionate."

DB: If we say compassion is the same as sodium, it is universal. Then every person's compassion is the same.

JK: And compassion, love, and intelligence. You can't be compassionate without intelligence.

DB: So we say intelligence is universal too!

JK: Obviously.

DB: But we have methods of testing intelligence in particular people, you see.

JK: Oh, no.

DB: But perhaps that is all part of the thing that is getting in the way?

JK: Part of this divisive, fragmentary way of thinking.

DB: Well, there may be holistic thinking, although we are not in it yet.

JK: Then holistic thinking is not thinking; it is some other factor.

DB: Some other factor that we haven't gone into yet.

JK: If love is common to all of us, why am I blind to it?

DB: I think partly because the mind boggles; it just refuses to consider such a fantastic change of concept, of a way of looking.

JK: But you said just now that sodium is sodium.

DB: You see, we have a lot of evidence for that in all sorts of experiments, built up through a lot of work and experience. Now, we can't do that with love. You can't go into a laboratory and prove that love is love.

JK: Oh, no. Love isn't knowledge. But why does one's mind refuse to accept a very obvious factor? Is it the fear of letting go my old values, standards, and opinions?

DB: I think it is probably something deeper. It is hard to pin down, but it isn't any of those simple things, although what you suggest is a partial explanation.

JK: That is a superficial explanation, I know. Is it the deep rooted anxiety, the longing to be totally secure?

DB: But that again is based on fragmentation.

JK: Of course.

DB: If we accept that we are fragmented, we will inevitably want to be totally secure, because being fragmented, we are always in danger.

JK: Is that the root of it? This urge, this demand, this longing to be totally secure in our relationship with everything? To be certain? Of course, in nothingness there is complete security!

DB: It is not the demand for security which is wrong, but the demand that the fragment be secure, which it cannot possibly be.

JK: That is right. Like each country trying to be secure, it is not secure.

DB: But complete security could be achieved if all the countries got together. The way you have put it sounds as if we should live eternally in insecurity, you see.

JK: No, we have made that very clear.

DB: It makes sense to ask for security, but we are going about it the wrong way.

JK: Yes, that's right. So how do we convey that love is universal, not personal, to a man who has lived completely in the narrow groove of personal achievement?

DB: It seems the first point is will he question his narrow, "unique" personality?

JK: People question it; they see the logic of what we are discussing, yet, curiously, people who are very serious in these matters have tried to find the wholeness of life through starvation, through torture—you know, every kind of way. But you can't apprehend or perceive or be the whole through torture, discipline, all that. So what shall we do? Let's say I have a brother who refuses to see all this. And as I have great affection for him, I want him to move out of this. And I have tried to communicate with him verbally and sometimes nonverbally, by a gesture or by a look, but all this is still from the outside. And perhaps that is the reason why he resists. Can I point out to my brother that in himself this flame can be awakened? It means he must listen to me, but, back again, my brother refuses to listen!

DB: It seems that there are some actions which are not possible. If a person is caught in a certain thought such as fragmentation, then he can't change it, because there are a lot of other thoughts behind it. So we have to find a place where he is free to act, to move, which is not controlled by the conditioning.

JK: Of course.

DB: Thoughts he doesn't know. He is not actually free to take this action because of the whole structure of thought that holds him.

JK: So how do I help—I use that word with great caution—my brother? What is the root of all this? We said becoming, but all that is verbal; it can be explained in ten different ways—the cause, the effect, and all the rest of it. After I explain all this, he says, "You have left me where I am." And my intelligence, my affection, love says, "I can't let him go." I can't say, "Well, go to hell," and move on. Which means am I putting pressure on him?

I am not using any kind of pressure or reward, none of that. My responsibility is that I can't let another human being go. It is not the responsibility of duty and all that dreadful stuff. But it is the responsibility of intelligence to say all that. [*Pause*] There is a tradition in India that one who is called the Maitreya Buddha took a vow that he would not become the ultimate Buddha until he had liberated other human beings too.

DB: Altogether?

JK: Yes. You see, that tradition hasn't changed anything. How can one? If one has that intelligence, that compassion, that love—which is not of a country, a person, an ideal, or a saviour—if one has the purity of that, can that be transmitted to another? By living with him, talking to him? You see, it can all become mechanical.

DB: Would you say that this question has never really been solved?

JK: I think so. But we must solve it. You follow? It has not been solved, but our intelligence says solve it. No, I think intelligence doesn't say solve it; intelligence says these are the facts and perhaps some will capture it.

DB: Well, it seems to me that there are really two factors: One is the preparation by reason to show that it all makes sense, and from there possibly some will capture it.

JK: We have done that, sir. The map has been laid out, and I have seen it very clearly: the conflicts, the misery, the confusion, the insecurity, the becoming. All that is extremely clear. But at the end of the chapter I am back at the beginning. Or perhaps I have got a glimpse of it, and my craving is to capture that glimpse and hold on to it. And that becomes a memory. You follow? And all the nightmare begins!

In your showing me the map very clearly, you have also pointed out to me something much deeper than that, which is love. And by your person, your reasoning, your logic, I am groping, seeking after that. But the weight of my body, my brain, my tradition—all that draws me back. So it is a constant battle—and I think the whole thing is so wrong.

DB: What is wrong?

JK: The way we are living.

DB: Many people must see that by now.

JK: We have asked whether man has taken a wrong turning and entered into a valley where there is no escape. That can't be so; that is too depressing, too appalling.

DB: I think some people might object to that. The very fact that it is appalling does not make it untrue. I think you would have to give some stronger reason why you feel that to be untrue.

JK: Oh, yes.

DB: Do you perceive in human nature some possibility of a real change?

JK: Of course, sir. Otherwise . . .

DB: It would be meaningless.

JK: We'd be monkeys, machines! You see, the faculty for radical change is attributed to some outside agency, and therefore we look to that and get lost in that. If we don't look to anybody and are completely free from dependence, then that solitude is common to all of us. It is not an isolation. It is an obvious fact that when you see all this—this is so ugly, unreal, so stupid—you are naturally solitary, you are naturally alone. That sense of aloneness is common.

DB: Yes, but in the ordinary sense of loneliness each person feels it is his own.

JK: Loneliness is not solitude; it is not aloneness.

DB: I think all the fundamental things are universal, and therefore you are saying that when the mind goes deep, it comes into something universal.

JK: That's right.

DB: Whether or not you call it absolute.

JK: The problem is to make the mind go very, very deeply into itself.

DB: Yes. Now, there is something that has occurred to me. When we start with a particular problem, our mind is very shallow, then we go to something more general. The word "general" has the same root as "to generate"; the genus is the coming generation . . .

JK: To generate, of course.

DB: When we go to something more general, a depth is generated. But going on, still further, the general is still limited because it is thought.

JK: Quite right. But to go so profoundly requires not only tremendous courage but the sense of constantly pursuing the same stream.

DB: Well, one might call that "diligence"; but that is still too limited, right?

JK: Yes, "diligence" is too limited. It goes with a religious mind in a sense that it is diligent in its action, its thoughts, and so on, but it is still limited. If the mind can go from the particular to the general and from the general . . .

DB: . . . to the absolute, to the universal. But many people would say that is all very abstract and has nothing to do with daily life.

JK: I know. Yet it is the most practical thing and not an abstraction.

DB: In fact, it is the particular that is the abstraction.

JK: Absolutely. The particular is the most dangerous.

DB: It is also the most abstract, because you only get to the particular by abstracting from the whole.

JK: Of course, of course.

DB: But I think this may be part of the problem. People feel they want something that really affects us in daily life. They don't just want to get lost in talking; therefore they say, "All these vaporous generalities don't interest us, and we are getting into the real, solid concrete facts of daily life." And it is true that what we are discussing must work in daily life, but daily life does not contain the solution of its problems.

JK: No, the daily life is the general and the particular.

DB: The human problems which arise in daily life cannot be solved there.

JK: From the particular move to the general. From the general move still deeper, and there perhaps is the purity of what is called compassion, love, and intelligence. But that means giving your mind to this, your heart; your whole being must be involved in this.

We have talked now for a long time. Have we reached somewhere?

DB: Possibly so, yes.

JK: I think so.

The Future
of Humanity

Introduction *by* David Bohm

These two dialogues took place three years after a series of similar dialogues between Krishnamurti and myself, which appeared in the book *The Ending of Time* [Harper and Row, 1985]. Therefore they were inevitably profoundly affected by what had been done in these earlier dialogues. In a certain sense, therefore, both sets of dialogues deal with closely related questions. Of course, *The Ending of Time* can, because of its much greater length, go into these questions in a more thorough and extensive way. Nevertheless, these two dialogues stand by themselves; they approach the problems of human life in their own way and provide important additional insights into these problems. Moreover, I feel they are easier to follow and may therefore usefully serve as an introduction to *The Ending of Time*.

The starting point for our discussions was the question "What is the future of humanity?" This question is by now of vital concern to everyone, because modern science and technology are clearly seen to have opened up immense possibilities of destruction. It soon became clear as we talked together that the ultimate origin of this situation is in the generally confused mentality of mankind, which has not changed basically in this respect throughout the whole of recorded history and probably for much longer than this. Evidently, it was essential to enquire deeply into the root of this difficulty if there is ever to be a possibility that humanity will be diverted from its present very dangerous course.

These dialogues constitute a serious enquiry into this problem, and as they proceeded, many of the basic points of Krishnamurti's teachings emerged. Thus, the question of the future of humanity seems, at first sight, to imply that a solution must involve time in a fundamental way. Yet, as Krishnamurti points out, psychological time, or "becoming," is the very source of the destructive current that is putting the future of humanity at risk. To question time in this way, however, is to question the adequacy of knowledge and thought, as a means of dealing with this problem. But if knowledge and thought are not adequate, what is it that is actually required? This led in turn to the question of whether mind is limited by the brain of mankind, with all the knowledge that it has accumulated over the ages. This knowledge, which now conditions us deeply, has produced what is, in effect, an irrational and self-destructive programme in which the brain seems to be helplessly caught up.

If mind is limited by such a state of the brain, then the future of humanity must be very grim indeed. Krishnamurti does not, however, regard these limitations as inevitable. Rather, he emphasizes that mind is essentially free of the distorting bias that is inherent in the conditioning of the brain, and that, through insight arising in proper undirected attention without a centre, it can change the cells of the brain and remove the destructive conditioning. If this is so, then it is crucially important that there be this kind of attention and that we give to this question the same intensity of energy that we generally give to other activities of life that are really of vital interest to us.

At this point, it is worth remarking that modern research into the brain and nervous system actually gives considerable support to Krishnamurti's statement that insight may change the brain cells. Thus, for example, it is now well known that there are important substances in the body, the hormones and the neurotransmitters, that fundamentally affect the entire functioning of the brain and nervous system. These substances respond, from moment to moment, to what a person knows, to what he thinks, and to what all this means to him. It is by now fairly well established that in this way the brain cells and their functioning are profoundly affected by knowledge and thought, especially when these give rise to strong feelings and passions. It is thus quite plausible that insight, which must arise in a state of great mental energy and passion, could change the brain cells in an even more profound way.

What has been said here necessarily gives only a brief outline of what is in the dialogues and cannot show the full scope and depth of the enquiry that takes place within them into the nature of human consciousness and of the problems that have arisen in this consciousness. Indeed, I would say that they are both concise and easily readable, containing the essential spirit of the whole of Krishnamurti's teachings and throwing an important further light on them.

—DAVID BOHM, 1986

Is There an Action
Not Touched by Thought?

11 JUNE 1983, BROCKWOOD PARK, HAMPSHIRE

JIDDU KRISHNAMURTI: I thought we would talk about the future of man, about humanity.

DAVID BOHM: The whole of mankind.

JK: Not just the British or the French or the Russians or the Americans, but human beings as a whole.

DB: The future is all interlinked now anyway.

JK: As things are, from what one observes, the world has become tremendously dangerous.

DB: Yes.

JK: Terrorists, wars, and the national and racial divisions, some dictators who want to destroy the world, and so on.

DB: Yes, and there is the economic crisis and the ecological crisis...

JK: Yes, ecological and economic problems—problems seem to multiply more and more. So what is the future of man? What is the future not only of the present generation but of the coming generations?

DB: Yes, well, the future looks very grim.

JK: Very grim. If you were quite young and I was quite young, what would we do, knowing all this? What would be our reaction, what would be our life, our way of earning a livelihood?

DB: Yes, I've often thought of that. For example, I've asked myself would I go into science again. And I'm not at all certain now, because science does not seem to be relevant to this crisis.

JK: No, on the contrary, scientists are helping.

DB: That makes it worse. They might help in a right way, but in fact . . .

JK: So what would you do? I think I would stick to what I'm doing.

DB: Well, that would be easy for you.

JK: For me it would be easy.

DB: There are several problems that we might discuss. One is when a person is just starting out, he has to make a living. There are very few opportunities now, and most of these are in jobs which are extremely limited.

JK: And there is unemployment throughout the world. I wonder what he would do, knowing that the future is grim, very depressing, dangerous, and so uncertain. Where would you begin?

DB: Well, I think one would have to stand back from all the particular problems of one's own needs and the needs of people around one.

JK: Are you saying one should really forget oneself for the time being?

DB: Yes.

JK: Even if I did forget myself, when I look at this world in which I am going to live, and have some kind of career or profession, what would I do? This is a problem that I think most young people are facing.

DB: Yes. That's clear. Well, have you something that you would suggest?

JK: You see, I don't think in terms of evolution.

DB: I understand that. That's the point I was expecting we would discuss.

JK: I don't think there is psychological evolution at all.

DB: We have discussed this quite often, so I think I understand to some extent what you mean. But I think that people who are new to this are not going to understand.

JK: Yes, we will discuss this whole question, if you will. Why are we concerned about the future? Surely the whole future is now.

DB: In a sense the whole future is now, but we have to make that clear. It goes very much against the whole way of thinking, of the tradition of mankind.

JK: I know. Mankind thinks in terms of evolution, continuance, and so on.

DB: Perhaps we could approach it in another way? That is, evolution seems in the present era to be the most natural way to think. So I would like to ask you what objections you have to thinking in terms of evolution. Could I explain a point? This word "evolution" has many meanings.

JK: Of course. We are talking psychologically.

DB: Now, the first point is let's dispose of it physically.

JK: An acorn will grow into an oak.

DB: Also the species have evolved: for example, from the plants to the animals and to man.

JK: Yes, we have taken a million years to be what we are.

DB: You have no question that that has happened?

JK: No, that has happened.

DB: It may continue to happen.

JK: That is evolution.

DB: That is a valid process.

JK: Of course.

DB: It takes place in time. And therefore, in that region, the past, present, and future are important.

JK: Yes, obviously. I don't know a certain language; I need time to learn it.

DB: Also it takes time to improve the brain. You see, if the brain started out small and then it got bigger and bigger, that took a million years.

JK: And it becomes much more complex and so on. All that needs time. All that is movement in space and time.

DB: Yes. So you will admit physical time and neurophysiological time.

JK: Neurophysiological time, absolutely. Of course. Any sane man would.

DB: Most people also admit psychological time, what they call mental time.

JK: Yes, that is what we are talking about. Whether there is such a thing as psychological tomorrow, psychological evolution.

DB: Or yesterday. Now, at first sight I am afraid this will sound strange. It seems I can remember yesterday. And there is tomorrow; I can anticipate. And it has happened many times; you know days have succeeded each other. So I do have the experience of time, from yesterday to today to tomorrow.

JK: Of course. That is simple enough.

DB: Now, what is it that you are denying?

JK: I deny that I will be something, become better.

DB: I can change . . . but now there are two ways of looking at that. One approach is will I intentionally become better because I am trying? Or is evolution a natural, inevitable process, in which we are being swept along as if in a current and perhaps becoming better, or worse, or finding that something is happening to us.

JK: Psychologically.

DB: Psychologically, which takes time, which may not be the result of my trying to become better. It may or may not be. Some people think one way, some another. But are you denying also that there is a kind of natural psychological evolution as there was a natural biological evolution?

JK: I am denying that, yes.

DB: Now, why do you deny it?

JK: Because, first of all, what is the psyche, the me, the ego, and so on? What is it?

DB: The word "psyche" has many meanings. It may mean the mind, for example. Do you mean that the ego is the same thing?

JK: The ego. I am talking of the ego, the me.

DB: Yes. Now, some people think there will be an evolution in which the me is transcended, that it will rise to a higher level.

JK: Yes. Will the transition need time?

DB: A transcendence, a transition.

JK: Yes. That is my whole question.

DB: So there are two questions. One is will the me ever improve? And the other is even if we suppose we want to get beyond the me, can that be done in time?

JK: That cannot be done in time.

DB: Now, we have to make it clear why not.

JK: Yes. I will. We will go into it. What is the me? If the psyche has such different meanings, the me is the whole movement which thought has brought about.

DB: Why do you say that?

JK: The me is the consciousness, my consciousness: The me is my name, form, and all the experiences, remembrances, and so on that I have had. The whole structure of the me is put together by thought.

DB: That again would be something which some people might find it hard to accept.

JK: Of course. We are discussing it.

DB: Now, the first experience, the first feeling I have about the me is that it is there independently and that the me is thinking.

JK: Is the me independent of my thinking?

DB: Well, my own first feeling is that the me is there independent of my thinking. And that it is the me that is thinking, you see.

JK: Yes.

DB: Just as I am here, and I could move; I could move my arm, I could think, or I could move my head. Now, is that an illusion?

JK: No.

DB: Why not?

JK: Because when I move my arm there is the intention to grasp something, to take something, which is first the movement of thought. That makes the arm move and so on. My contention is—

and I am ready to accept it as false or true—that thought is the basis of all this.

DB: Yes. Your contention is that the whole sense of the me and what it is doing is coming out of thought. Now, what you mean by thought is not merely intellectual?

JK: No, of course not. Thought is the movement of experience, knowledge, and memory. It is this whole movement.

DB: It sounds to me as if you mean the consciousness as a whole.

JK: As a whole—that's right.

DB: And you are saying that that movement is the me?

JK: The whole content of that consciousness is the me. That me is not different from my consciousness.

DB: Yes. I think one could say that I am my consciousness, for if I am not conscious, I am not here.

JK: Of course.

DB: Now, is consciousness nothing but what you have just described, which includes thought, feeling, intention?

JK: Intention, aspirations . . .

DB: Memories . . .

JK: Memories, beliefs, dogmas, the rituals that are performed. The whole, like the computer that has been programmed.

DB: Yes. Now that certainly is in consciousness. Everybody would agree, but many people would feel that there is more to it than that; that consciousness may go beyond that.

JK: Let's go into it. The content of our consciousness makes up the consciousness.

DB: Yes, I think that requires some understanding. The ordinary use of the word "content" is quite different. If you say that the content of a glass is water, the glass is one thing and the water is another.

JK: Consciousness is made up of all that it has remembered: beliefs, dogmas, rituals, fears, pleasures, sorrow.

DB: Yes. Now, if all that were absent, would there be no consciousness?

JK: Not as we know it.

DB: But there would still be a kind of consciousness?

JK: A totally different kind. But consciousness, as we know it, is all that.

DB: As we generally know it.

JK: Yes. And that is the result of multiple activities of thought. Thought has put all this together, which is my consciousness—the reactions, the responses, the memories; extraordinary, complex intricacies, and subtleties. All that makes up consciousness.

DB: As we know it.

JK: But does that consciousness have a future?

DB: Yes. Does it have a past?

JK: Of course. Remembrance.

DB: Remembrance, yes. Why do you ask if it has a future then?

JK: If it has a future, it will be exactly the same kind of thing, moving. The same activities, the same thoughts, modified, but the pattern will be repeated over and over again.

DB: Are you saying that thought can only repeat?

JK: Yes.

DB: But there is a feeling, for example, that thought can develop new ideas.

JK: But thought is limited because knowledge is limited.

DB: Well, yes, that might require some discussion.

JK: Yes, we must discuss it.

DB: Why do you say knowledge is always limited?

JK: Because you, as a scientist, are experimenting, adding, searching. And after you, some other person will add more. So knowledge, which is born of experience, is limited.

DB: But some people have said it isn't. They would hope to obtain perfect, or absolute, knowledge of the laws of nature.

JK: The laws of nature are not the laws of human beings.

DB: Well, do you want to restrict the discussion then to knowledge about the human being?

JK: Of course, that's all we can talk about.

DB: Even there it is a question of whether that knowledge of nature is possible too.

JK: Yes. We are talking about the future of humanity.

DB: So are we saying that man cannot obtain unlimited knowledge of the psyche?

JK: That's right.

DB: There is always more that is unknown.

JK: Yes. There is always more and more unknown. So if once we admit that knowledge is limited, then thought is limited.

DB: Yes, thought depends on knowledge, and the knowledge does not cover everything. Therefore thought will not be able to handle everything that happens.

JK: That's right. But that is what the politicians and all the other people are doing. They think thought can solve every problem.

DB: Yes. You can see in the case of politicians that knowledge is very limited; in fact it is almost nonexistent! And, therefore, when you lack adequate knowledge of what you are dealing with, you create confusion.

JK: Yes. So then as thought is limited, our consciousness, which has been put together by thought, is limited.

DB: Now, can you make that clear? That means we can only stay in the same circle.

JK: The same circle.

DB: You see, one of the ideas might be, if you compare with science, that people might think that although their knowledge is limited, they are constantly discovering.

JK: What you discover is added to but is still limited.

DB: It is still limited. That's the point. I can keep on; I think one of the ideas behind a scientific approach is that though knowledge is limited, I can discover and keep up with the actuality.

JK: But that is also limited.

DB: My discoveries are limited. And there is always the unknown which I have not discovered.

JK: That is what I am saying. The unknown, the limitless, cannot be captured by thought.

DB: Yes.

JK: Because thought in itself is limited. You and I agree to that; we not only agree but it is a fact.

DB: Perhaps we could bring it out still more. That is, thought is limited, even though we might intellectually consider that thought is not limited. There is a very strong predisposition, tendency, to feel that way—that thought can do anything.

JK: Anything. It can't. See what it has done in the world.

DB: Well, I agree that it has done some terrible things, but that doesn't prove that it is always wrong. You see, perhaps you could blame it on the people who have used it wrongly.

JK: I know. That is a good old trick! But thought in itself is limited; therefore whatever it does is limited.

DB: Yes, and you are saying that it is limited in a very serious way.

JK: That's right. Of course in a very, very serious way.

DB: Could we bring that out? Say what that way is?

JK: That way is what is happening in the world.

DB: All right, let's look at that.

JK: The totalitarian ideals are the invention of thought.

DB: The very word "totalitarian" means that people wanted to cover the totality, but they couldn't.

JK: They couldn't.

DB: The thing collapsed.

JK: It is collapsing.

DB: But then there are those who say they are not totalitarians.

JK: But the democrats, the republicans, the idealists, and so on, all their thinking is limited.

DB: Yes, and it is limited in a way that is . . .

JK: . . . very destructive.

DB: Now, could we bring that out? You see, I could say, "All right, my thought is limited, but it may not be all that serious." Why is this so important?

JK: That is fairly simple: because whatever action is born of limited thought must inevitably breed conflict.

DB: Yes.

JK: Like dividing humanity religiously, or into nationalities and so on, has created havoc in the world.

DB: Yes. Now, let's connect that with the limitation thought. My knowledge is limited: How does that lead me to divide the world into . . .

JK: Aren't we seeking security?

DB: Yes.

JK: And we thought there was security in the family, in the tribe, in nationalism. So we thought there was security in division.

DB: Yes. Now, it has come out. Take the tribe, for example: One may feel insecure, and one then says, "With the tribe I am secure." That is a conclusion. And I think I know enough to be sure that is so—but I don't. Other things happen that I don't know, which make that very insecure. Other tribes come along.

JK: No, no! The very division creates insecurity.

DB: Yes, it helps to create it, but I am trying to say that I don't know enough to know that. I don't see that.

JK: But one doesn't see it because one has not thought about anything, not looked at the world, as a whole.

DB: Well, the thought which aims at security attempts to know everything important. As soon as it knows everything important, it says, "This will bring security." But there are a lot of things it still doesn't know, and one is that this very thought itself is divisive.

JK: Yes. In itself, it is limited. Anything that is limited must inevitably create conflict. If I say I am an individual, that is limited.

DB: Yes.

JK: I am concerned with myself; that is very limited.

DB: We have to make this clear. If I say this is a table which is limited, it creates no conflict.

JK: No, there is no conflict there.

DB: But when I say this is "me," that creates conflict.

JK: The "me" is a divisive entity.

DB: Let us see more clearly why

JK: Because it is separative; it is concerned with itself. The "me" identifying with the greater nation is still divisive.

DB: I define myself in the interest of security so that I know what I am as opposed to what you are, and I protect myself. Now, this creates a division between me and you.

JK: We and they and so on.

DB: That comes from my limited thought, because I don't understand that we are really closely related and connected.

JK: We are human beings, and all human beings have more or less the same problems.

DB: But I haven't understood that. My knowledge is limited and I think that we can make a distinction and protect ourselves, and me, and not the others.

JK: Yes, that's right.

DB: But in the very act of doing that I create instability.

JK: That's right, insecurity. So if not merely intellectually or verbally but actually we feel that we are the rest of humanity, then the responsibility becomes immense.

DB: Well, how can you do anything about that responsibility?

JK: Then I either contribute to the whole mess or keep out of it. That means to be at peace, to have order in oneself—I will come to that; I am going too fast.

DB: Well, I think we have touched upon an important point. We say the whole of humanity, of mankind, is one, and therefore to create division is ...

JK: ... dangerous.

DB: Yes. Whereas to create division between me and the table is not dangerous, because in some sense we are not one.

JK: Of course.

DB: That is, only in some very general sense are we at one. Now, mankind doesn't realize that it is all one.

JK: Why?

DB: Let's go into it. This is a crucial point. There are so many divisions, not only between nations and religions but between one person and another.

JK: Why is there this division?

DB: The feeling is, at least in the modern era, that every human being is an individual. This may not have been so strong in the past.

JK: That is what I question. I question altogether whether we are individuals.

DB: That is a big question ...

JK: Of course. We said just now that the consciousness which is me is similar to the rest of mankind. They all suffer, all have fears, are insecure; they have their own particular gods and rituals, all put together by thought.

DB: I think there are two questions here. One is not everybody feels that he is similar to others. Most people feel they have some unique distinction.

JK: What do you mean by "unique distinction"? Distinction in doing something?

DB: There may be many things. For example, one nation may feel that it is able to do certain things better than another; one person has some special things he does or a particular quality.

JK: Of course. Somebody else is better in this or that.

DB: He may take pride in his own special abilities or advantages.

JK: But when you put away that, basically we are the same.

DB: You are saying these things which you have just described are . . .

JK: . . . superficial.

DB: Yes. Now, what are the things that are basic?

JK: Fear, sorrow, pain, anxiety, loneliness, and all the human travail.

DB: But many people might feel that the basic things are the highest achievements of man. For one thing, people may feel proud of man's achievement in science and art and culture and technology.

JK: We have achieved in all those directions, certainly. In technology, communication, travel, medicine, surgery, we have advanced tremendously.

DB: Yes, it is really remarkable in many ways.

JK: There is no question about it. But what have we psychologically achieved?

DB: None of this has affected us psychologically.

JK: Yes, that's right.

DB: And the psychological question is more important than any of the others, because if the psychological question is not cleared up, the rest is dangerous.

JK: Yes. If we are psychologically limited, then whatever we do will be limited, and the technology will then be used by our limited...

DB: Yes, the master is this limited psyche and not the rational structure of technology. And in fact technology then becomes a dangerous instrument. So that is one point, that the psyche is at the core of it all, and if the psyche is not in order, then the rest is useless. Then, although we are saying there are certain basic disorders in the psyche common to us all, we may all have a potential for something else. The next point is are we all one really? Even though we are all similar that doesn't mean we are all the same, that we are one.

JK: We said in our consciousness basically we all have the same ground on which we stand.

DB: Yes, but the fact that the human body is similar doesn't prove we are all the same.

JK: Of course not. Your body is different from mine.

DB: Yes, we are in different places, we are different entities, and so on. But I think you are saying that the consciousness is not an entity which is individual.

JK: That's right.

DB: The body is an entity which has a certain individuality.

JK: That all seems so clear. Your body is different from mine. I have a different name from you.

DB: Yes, we are different. Though of similar material, we are different. We can't exchange because the proteins in one body may not agree with those in the other. Now, many people feel that way about the mind, saying that there is a chemistry between people which may agree or disagree.

JK: Yes, but actually if you go deeper into the question, consciousness is shared by all human beings.

DB: But the feeling is that the consciousness is individual and that it is communicated.

JK: I think that is an illusion, because we are sticking to something that is not true.

DB: Do you want to say that there is one consciousness of mankind?

JK: It is all one.

DB: That is important, because whether it is many or one is a crucial question.

JK: Yes.

DB: It could be many, which are then communicating and building up the larger unit. Or are you saying that from the very beginning it is all one?

JK: From the very beginning it is all one.

DB: And the sense of separateness is an illusion?

JK: That is what I am saying, over and over again. That seems so logical, sane. The other is insanity.

DB: Yes, but people don't feel, at least not immediately, that the notion of separate existence is insane, because one extrapolates from the body to the mind. One says, it is quite sensible to say my body is separate from yours, and inside my body is my mind. Now, are you saying that the mind is not inside the body?

JK: That is quite a different question. Let's finish with the other first. Each one of us thinks that we are separate individuals psychically. And what we have done in the world is a colossal mess.

DB: Well, if we think we are separate when we are not separate, then it will clearly be a colossal mess.

JK: That is what is happening. Each one thinks he has to do what he wants to do; fulfil himself. So he is struggling in his separateness to achieve peace, to achieve security, and that security and peace are totally denied.

DB: The reason they are denied is because there is no separation. You see, if there were really separation it would be a rational thing to try to do. But if we are trying to separate what is inseparable, the result will be chaos.

JK: That's right.

DB: Now, that is clear, but I think that it will not be clear to people immediately that the consciousness of mankind is one inseparable whole.

JK: Yes, an inseparable whole.

DB: Many questions will arise if we consider this notion, but I don't know if we have gone far enough into it yet. One question is why do we think we are separate?

JK: Why do I think I am separate? That is my conditioning.

DB: Yes, but how did we ever adopt such a foolish conditioning?

JK: From childhood, it is mine, my toy, not yours.

DB: But the first feeling I get of "it is mine" is because I feel I am separate. It isn't clear how the mind, which was one, came to this illusion that it is all broken up into many pieces.

JK: I think it is again the activity of thought. Thought in its very nature is divisive, fragmentary, and therefore I am a fragment.

DB: Thought will create a sense of fragments. You could see, for example, that once we decide to set up a nation, we will think we are separate from other nations, and all sorts of consequences follow, which make the whole thing seem independently real. We have a separate language, a separate flag, and we set up a boundary. And after a while we see so much evidence of separation that we forget how it started and say it was there always, and that we are merely proceeding from what was there always.

JK: Of course. That's why I feel that if once we grasp the nature and structure of thought, how thought operates, what is the source of thought—and therefore it is always limited—if we really see that, then . . .

DB: Now, the source of thought is what? Is it memory?

JK: Memory. The remembrance of things past, which is knowledge, and knowledge is the outcome of experience, and experience is always limited.

DB: Well, thought also includes, of course, the attempt to go forward, to use logic, to take into account discoveries and insights.

JK: As we were saying some time ago, thought is time.

DB: All right. Thought is time. That requires more discussion too, because the first response is to say time is there first, and thought is taking place in time.

JK: Ah, no.

DB: For example, if movement is taking place, if the body is moving, this requires time.

JK: To go from here to there needs time. To learn a language needs time.

DB: Yes. To grow a plant needs time.

JK: To paint a picture takes time.

DB: We also say that to think takes time.

JK: So we think in terms of time.

DB: Yes, the first point that one would tend to look at is whether just as everything takes time, to think takes time, right? You are saying something else, which is that thought is time.

JK: Thought is time.

DB: That is psychologically speaking.

JK: Psychologically, of course.

DB: Now, how do we understand that?

JK: How do we understand what?

DB: That thought is time. You see, it is not obvious.

JK: Oh, yes. Would you say thought is movement, and time is movement?

DB: That is movement. You see, time is a mysterious thing; people have argued about it. We could say that time requires movement. I could understand that we cannot have time without movement.

JK: Time is movement. Time is not separate from movement.

DB: I don't say it is separate from movement. You see, if we said time and movement are one . . .

JK: Yes, we are saying that.

DB: They cannot be separated, right?

JK: No.

DB: That seems fairly clear. Now, there is physical movement, which means physical time. There's the heartbeat and so on.

JK: Physical time, hot and cold, and also dark and light . . .

DB: The seasons . . .

JK: Sunset and sunrise. All that.

DB: Yes. Now then, we have the movement of thought. That brings in the question of the nature of thought. Is thought nothing but a movement in the nervous system, in the brain? Would you say that?

JK: Yes.

DB: Some people have said it includes the movement of the nervous system, but there might be something beyond.

JK: What is time, actually? Time is hope.

DB: Psychologically.

JK: Psychologically. I am talking entirely psychologically for the moment. Hope is time. Becoming is time. Achieving is time. Now, take the question of becoming. I want to become something, psychologically. I want to become nonviolent. Take that, for example. That is altogether a fallacy.

DB: We understand it is a fallacy, but the reason it is a fallacy is that there is no time of that kind. Is that it?

JK: No. Human beings are violent.

DB: Yes.

JK: And they have been talking a great deal—Tolstoy and in India—of nonviolence. The fact is we are violent. And the nonviolence is not real. But we want to become that.

DB: But it is again an extension of the kind of thought that we have with regard to material things. If you see a desert, the desert is real and you say the garden is not real, but in your mind is the garden which will come when you put the water there. So we say we can plan for the future when the desert will become fertile. Now, we have to be careful, we say we are violent, but we cannot by similar planning become nonviolent.

JK: No.

DB: Why is that?

JK: Why? Because the nonviolent state cannot exist when there is violence. That's just an ideal.

DB: One has to make this more clear, in the same sense the fertile state and the desert don't exist together either. I think you are saying that in the case of the mind, when you are violent, nonviolence has no meaning.

JK: Violence is the only state.

DB: That is all there is.

JK: Yes, not the other.

DB: The movement towards the other is illusory.

JK: So all ideals are illusory, psychologically. The ideal of building a marvellous bridge is not illusory. You can plan it, but to have psychological ideals . . .

DB: Yes, if you are violent and you continue to be violent while you are trying to be nonviolent, it has no meaning.

JK: No meaning, and yet that has become such an important thing. The becoming, which is either becoming "what is" or becoming away from "what is."

DB: Yes. "What should be." If you say there can be no sense to becoming in the way of self-improvement, that's . . .

JK: Oh, self-improvement is something so utterly ugly. We are saying that the source of all this is a movement of thought as time. When once we admit time, psychologically, all the other ideals—nonviolence, achieving some super state, and so on—become utterly illusory.

DB: Yes. Now, when you talk of the movement of thought as time, it seems to me that that time which comes from the movement of thought is illusory.

JK: Yes.

DB: We sense it as time, but it is not a real kind of time.

JK: That is why we asked, what is time?

DB: Yes.

JK: I need time to go from here to there. I need time if I want to learn engineering. I must study it; it takes time. That same movement is carried over into the psyche. We say, I need time to be good. I need time to be enlightened.

DB: Yes, that will always create a conflict. One part of you and another. So that movement in which you say I need time, also creates a division in the psyche. Between the observer and the observed.

JK: Yes. We are saying the observer is the observed.

DB: And therefore there is no time, psychologically.

JK: That's right. The experiencer, the thinker, is the thought. There is no thinker separate from thought.

DB: All that you are saying seems very reasonable, but I think that it goes so strongly against the tradition we are used to that it will be extraordinarily hard for people, generally speaking, really to understand.

JK: Most people just want a comfortable way of living: "Let me carry on as I am, for God's sake. Leave me alone!"

DB: But that is the result of so much conflict, that people are worn out by it, I think.

JK: But in escaping from conflict or not resolving it, conflict exists, whether we like it or not. So that is the whole point: Is it possible to live a life without conflict?

DB: Yes, that is all implicit in what has been said. The source of conflict is thought, or knowledge, or the past.

JK: So then one asks, is it possible to transcend thought?

DB: Yes.

JK: Or is it possible to end knowledge? I am putting it psychologically.

DB: Yes. We say that knowledge of material objects and things like that, knowledge of science, will continue.

JK: Absolutely. That must continue.

DB: But what you call self-knowledge is what you are asking to end, isn't it?

JK: Yes.

DB: On the other hand people have said—even you have said—that self-knowledge is very important.

JK: Self-knowledge is important, but if I take time to understand myself, I will understand myself eventually by examination, analysis, by watching my whole relationship with others, and so on— all that involves time. And I say there is another way of looking at the whole thing without time. Which is when the observer is the observed.

DB: Yes.

JK: In that observation there is no time.

DB: Could we go into that further? I mean, for example, if you say there is no time, but still you feel that you can remember an hour ago you were someone else. Now, in what sense can we say that there is no time?

JK: Time is division. As thought is division. That is why thought is time.

DB: Time is a series of divisions of past, present, future.

JK: Thought is divisive. So time is thought. Or thought is time.

DB: It doesn't exactly follow from what you said.

JK: Let's go into it.

DB: Yes. You see, at first sight one would think that thought makes divisions of all kinds, with the ruler and with all kinds of things, and also divides up intervals of time: past, present, and future. Now, it doesn't follow, from just that, that thought is time.

JK: Look, we said time is movement.

DB: Yes.

JK: Thought is also a series of movements. So both are movements.

DB: Thought is a movement, we suppose, of the nervous system and . . .

JK: You see, it is a movement of becoming. I am talking psychologically.

DB: Psychologically. But whenever you think, something is also moving in the blood, in the nerves, and so on. Now, when you talk of a psychological movement, do you mean just a change of content?

JK: Change of content?

DB: Well, what is the movement? What is moving?

JK: Look, I am this, and I am attempting to become something else psychologically.

DB: So that movement is in the content of your thought.

JK: Yes.

DB: If you say, "I am this and I am attempting to become that," then I am in movement. At least, I feel I am in movement.

JK: Say, for instance, that I am greedy. Greed is a movement.

DB: What kind of a movement is it?

JK: To get what I want, to get more. It is a movement.

DB: All right.

JK: And I find that movement painful. Then I try not to be greedy.

DB: Yes.

JK: The attempt not to be greedy is a movement of time, is becoming.

DB: Yes, but even the greed was becoming.

JK: Of course. So is the real question, is it possible not to become, psychologically?

DB: It seems that would require that you should not be anything psychologically. As soon as you define yourself in any way, then . . .

JK: No, we will define it in a minute or two.

DB: I meant if I define myself as greedy, say that I am greedy, or I am this or I am that, then either I will want to become something else or to remain what I am.

JK: Now, can I remain what I am? Can I remain not with non-greed but with greed? Greed is not different from me; greed is me.

DB: The ordinary way of thinking is that I am here, and I could either be greedy or not greedy.

JK: Of course.

DB: These are attributes which I may or may not have.

JK: But the attributes are me.

DB: Now, that again goes very much against our common language and experience.

JK: All the qualities, the attributes, the virtues, the judgments, the conclusions, and opinions are me.

DB: It seems to me that this would have to be perceived immediately as obvious.

JK: That is the whole question. To perceive the totality of this whole movement instantly. Then we come to the point—it sounds a little odd, and perhaps a little crazy, but it is not: Is it possible to perceive without all the movement of memory? To perceive something directly without the word, without the reaction, without the memories entering into perception.

DB: That is a very big question, because memory has constantly entered perception. It would raise the question of what is going to stop memory from entering perception?

JK: Nothing can stop it. But if we see the reason, the rationality of the activity of memory which is limited—in the very perception that it is limited, we have moved out of it into another dimension.

DB: It seems to me that you have to perceive the whole of the limitation of memory.

JK: Yes, not one part.

DB: You can see in general that memory is limited, but there are many ways in which this is not obvious. For example, with many of our reactions this is not obvious; they may be memory, but we don't experience them as memory. Suppose I am becoming: I experience greed, and I have the urge to become less greedy. I can remember that I am greedy but think that this "me" is the one who remembers, not the other way around, not that memory creates "me," right?

JK: All this really comes down to whether humanity can live without conflict. It basically comes to that. Can we have peace on this earth? The activities of thought never bring it about.

DB: It seems clear from what has been said that the activity of thought cannot bring about peace: It inherently brings about conflict.

JK: Yes, if we once really see that, our whole activity would be totally different.

DB: But are you saying then that there is an activity which is not thought? Which is beyond thought?

JK: Yes.

DB: And which is not only beyond thought but which does not require the cooperation of thought? That it is possible for this to go on when thought is absent?

JK: That is the real point. We have often discussed this, whether there is anything beyond thought. Not something holy, sacred—we are not talking of that. We are asking, is there an activity which is not touched by thought? We are saying there is and that that activity is the highest form of intelligence.

DB: Yes. Now we have brought in intelligence.

JK: I know. I purposely brought it in! So intelligence is not the activity of cunning thought. There is intelligence to build a table.

DB: Well, intelligence can use thought, as you have often said. That is, thought can be the action of intelligence. Would you put it that way?

JK: Yes.

DB: Or it could be the action of memory?

JK: That's it. Either it is the action born of memory, and memory being limited, therefore thought is limited, and it has its own activity, which then brings about conflict.

DB: I think this would connect with what people are saying about computers. Every computer must eventually depend on some kind of memory which is put in, programmed. And that must be limited.

JK: Of course.

DB: Therefore when we operate from memory we are not very different from a computer. The other way round perhaps: The computer is not very different from us.

JK: I would say a Hindu has been programmed for the last five thousand years to be a Hindu; or, in this country, you have been programmed as British, or as a Catholic or a Protestant. So we are all programmed to a certain extent.

DB: Yes, but you are bringing in the notion of an intelligence which is free of the programme, which is creative, perhaps.

JK: Yes. That intelligence has nothing to do with memory and knowledge.

DB: It may act in memory and knowledge but it has nothing to do with it.

JK: That's right. I mean how do you find out whether it has any reality and is not just imagination and romantic nonsense? To come to that, one has to go into the whole question of suffering, whether there is an end to suffering. And as long as suffering and fear and the pursuit of pleasure exist, there cannot be love.

DB: There are many questions there. Suffering, pleasure, fear, anger, violence, and greed—all of those are the response of memory.

JK: Yes.

DB: They are nothing to do with intelligence.

JK: They are all part of thought and memory.

DB: And as long as they are going on it seems that intelligence cannot operate in thought, or through thought.

JK: That's right. So there must be freedom from suffering.

DB: Well, that is a very key point.

JK: That is really a very serious and deep question. Whether it is possible to end suffering, which is the ending of "me."

DB: Yes. It may seem repetitious, but the feeling is that I am there, and I either suffer or don't suffer. I either enjoy things or suffer. Now, I think you are saying that suffering arises from thought; it is thought.

JK: Identification. Attachment.

DB: So what is it that suffers? Memory may produce pleasure, and then when it doesn't work it produces the opposite of the feeling of pleasure: pain and suffering.

JK: Not only that. Suffering is much more complex, isn't it?

DB: Yes.

JK: What is suffering? The meaning of the word is to have pain, to have grief, to feel utterly lost, lonely.

DB: It seems to me that it is not only pain but a kind of total, very pervasive pain.

JK: And suffering is the loss of someone.

DB: Or the loss of something very important.

JK: Yes, of course. Loss of my wife, my son, brother, or whatever it is, and the desperate sense of loneliness.

DB: Or else just simply the fact that the whole world is going into such a state.

JK: Of course . . . all the wars.

DB: It makes everything meaningless, you see.

JK: What a lot of suffering wars have created. And wars have been going on for thousands of years. That is why I am saying we are carrying on with the same pattern of the last five thousand years or more.

DB: One can easily see that the violence and hatred in wars will interfere with intelligence.

JK: Obviously.

DB: But some people have felt that by going through suffering they become . . .

JK: . . . intelligent?

DB: Purified, like going through the crucible.

JK: I know. That through suffering you learn. That through suffering your ego vanishes, is dissolved.

DB: Yes, dissolved, refined.

JK: It is not. People have suffered immensely, how many wars, how many tears, and the destructive nature of governments. And unemployment, ignorance . . .

DB: Ignorance of disease, pain, everything. But what is suffering really? Why does it destroy intelligence, or prevent it? What is going on?

JK: Suffering is a shock. I suffer; I have pain; it is the essence of the "me."

DB: The difficulty with suffering is that it is the me that is there that is suffering.

JK: Yes.

DB: And this me is really being sorry for itself in some way.

JK: My suffering is different from your suffering.

DB: Yes, it isolates itself. It creates an illusion of some kind.

JK: We don't see that suffering is shared by all humanity.

DB: Yes, but suppose we do see it is shared by all humanity?

JK: Then I begin to question what suffering is. It is not my suffering.

DB: That is important. In order to understand the nature of suffering I have to get out of this idea that it is my suffering, because as long as I believe it is my suffering I have an illusory notion of the whole thing.

JK: And I can never end it.

DB: If you are dealing with an illusion, you can do nothing with it. You see why: We have to come back. Why is suffering the suffering of many? At first it seems that I feel pain in the tooth, or else I have a loss or something has happened to me, and the other person seems perfectly happy.

JK: Happy, yes. But also he is suffering in his own way.

DB: Yes. At the moment he doesn't see it, but he has his problems too.

JK: Suffering is common to all humanity.

DB: But the fact that it is common is not enough to make it all one.

JK: It is actual.

DB: Are you saying that the suffering of mankind is all one, inseparable?

JK: Yes, that is what I have been saying.

DB: As is the consciousness of man?

JK: Yes, that's right.

DB: That when anybody suffers, the whole of mankind is suffering.

JK: The whole point is we have suffered from the beginning of time, and we haven't solved it. We haven't ended suffering.

DB: But I think you have said that the reason we haven't solved it is because we are treating it as personal, or as in a small group . . . and that is an illusion.

JK: Yes.

DB: And any attempt to deal with an illusion cannot solve anything.

JK: Thought cannot solve anything psychologically.

DB: Because you can say that thought itself divides. Thought is limited and is unable to see that this suffering is all one. And in that way divides it up as mine and yours.

JK: That's right.

DB: And that creates illusion, which can only multiply suffering. Now, it seems to me that the statement that the suffering of mankind is one is inseparable from the statement that the consciousness of mankind is one.

JK: Suffering is part of our consciousness.

DB: But one doesn't get the feeling immediately that this suffering belongs to the whole of mankind, you see.

JK: The world is me: I am the world. But we have divided it up into the British earth and the French earth and all the rest of it!

DB: Do you mean, by the world, the physical world or the world of society?

JK: The world of society, primarily the psychological world.

DB: So we say the world of society, of human beings, is one, and when I say I am that world, what does it mean?

JK: The world is not different from me.

DB: The world and I are one. We are inseparable.

JK: Yes. And that is real meditation; you must feel this, not just as a verbal statement: It is an actuality. I am my brother's keeper.

DB: Many religions have said that.

JK: That is just a verbal statement and they don't keep it; they don't do it in their hearts.

DB: Perhaps some have done it, but in general it is not being done?

JK: I don't know if anybody has done it. We human beings haven't done it. Our religions actually have prevented it.

DB: Because of division? Every religion has its own beliefs and its own organization.

JK: Of course. Its own gods and its own saviours.

DB: Yes.

JK: So from that, is that intelligence actual? You understand my question? Or is it some kind of fanciful projection, hoping that it will solve our problems? It is not to me. It is an actuality. Because the ending of suffering means love.

DB: Before we go on, let's clear up a point about "me." You see, you said it is not to me. Now, in some sense it seems that you are still defining an individual. Is that right?

JK: Yes. I am using the word "I" as a means of communication.

DB: But what does it mean? In some way, let's say there are two people, let's say A and B.

JK: Yes.

DB: So A says it is not to me—that seems to create a division between A and B.

JK: That's right. But B creates the division.

DB: Why?

JK: What is the relationship between the two?

DB: B is creating the division by saying, "I am a separate person," but it may confuse B further when A says, "It's not that way to me," right?

JK: That is the whole point, isn't it, in relationship? You feel that you are not separate and that you really have this sense of love and compassion, and I haven't got it. I haven't even perceived or gone into this question. What is your relationship to me? You have a relationship with me, but I haven't any relationship with you.

DB: Well, I think one could say that the person who hasn't seen is almost living in a world of dreams, psychologically, and therefore the world of dreams is not related to the world of being awake.

JK: That's right.

DB: But the fellow who is awake can at least perhaps awaken the other fellow.

JK: You are awake; I am not. Then your relationship with me is very clear. But I have no relationship with you; I cannot have one. I insist on division, and you don't.

DB: Yes, we have to say that in some way the consciousness of mankind has divided itself. It is all one, but it has divided itself by thought. And that is why we are in this situation.

JK: That is why. All the problems that humanity has now psychologically, as well as in other ways, are the result of thought. And we are pursuing the same pattern of thought, and thought will never solve any of these problems. So there is another kind of instrument, which is intelligence.

DB: Well, that opens up an entirely different subject. And you mentioned love as well. And compassion.

JK: Without love and compassion there is no intelligence. And you cannot be compassionate if you are attached to some religion, if you are tied to a post like an animal . . .

DB: Yes, as soon as the self is threatened, then it cannot . . .

JK: You see, self hides behind . . .

DB: . . . other things, I mean noble ideals.

JK: Yes, it has immense capacity to hide itself. So what is the future of humanity? From what one observes, it is leading to destruction.

DB: That is the way it seems to be going.

JK: Very gloomy, grim, and dangerous. If one has children, what is their future? To enter into all this? And go through the misery of it all. So education becomes extraordinarily important. But now education is merely the accumulation of knowledge.

DB: Every instrument that man has invented, discovered, or developed has been turned toward destruction.

JK: Absolutely. They are destroying nature; there are very few tigers now.

DB: They are destroying forests and agricultural land.

JK: Nobody seems to care.

DB: Well, most people are just immersed in their plans to save themselves, but others have plans to save humanity. I think also there is a tendency toward despair, implicit in what is happening now, in that people don't think anything can be done.

JK: Yes. And if they think something can be done, they form little groups and little theories.

DB: There are those who are very confident in what they are doing.

JK: Most prime ministers are very confident. They don't know what they are doing really!

DB: Yes, but then most people haven't much confidence in what they are doing themselves.

JK: I know. And if someone has tremendous confidence, I accept that confidence and go with them. What then is the future of mankind, the future of humanity? I wonder if anybody is concerned with it? Or each person, each group, is only concerned with its own survival?

DB: I think the first concern almost always has been with survival in either the individual or the group. That has been the history of mankind.

JK: Therefore, perpetual wars, perpetual insecurity.

DB: Yes, but this, as you said, is the result of thought, which makes the mistake, on the basis of being incomplete, of identifying the self with the group and so on.

JK: You happen to listen to all this. You agree to all this, you see the truth of all this. Those in power will not even listen to you.

DB: No.

JK: They are creating more and more misery. The world is becoming more and more dangerous. What is the point of our seeing something to be true, and what effect has it?

DB: It seems to me that if we think in terms of the effects, we are bringing in the very thing which is behind the trouble: time! Then the response would be to get in quickly and do something to change the course of events.

JK: And therefore form a society, foundation, organization, and all the rest of it.

DB: But, you see, our mistake is to feel that we must think about something, although that thought is incomplete. We don't really know what is going on, and people have made theories about it, but they don't know.

JK: If that is the wrong question, then as a human being, who is mankind, what is my responsibility, apart from effect and all the rest of it?

DB: Yes, we can't look toward effects. But it is the same as with A and B, that A sees and B does not.

JK: Yes.

DB: Now, suppose A sees something and most of the rest of mankind does not. Then, it seems, one could say mankind is in a sense dreaming, asleep.

JK: It is caught in illusion.

DB: Illusion. And the point is that if somebody sees something, his responsibility is to help awaken the others out of the illusions.

JK: That is just it. This has been the problem. That is why the Buddhists have projected the idea of the bodhisattva, who is the essence of all compassion and is waiting to save humanity. It sounds nice. It is a happy feeling that there is somebody doing this. But in actuality we won't do anything that is not comfortable, satisfying, secure, both psychologically and physically.

DB: That is basically the source of the illusion.

JK: How does one make others see all this? They haven't time, they haven't the energy, they haven't even the inclination. They want to be amused. How does one make X see this whole thing so clearly that he says, "All right, I have got it. I will work. And I see I am responsible," and all the rest of it. I think that is the tragedy of those who see and those who don't.

Second Conversation

Is There Evolution
of Consciousness?

20 JUNE 1983, BROCKWOOD PARK, HAMPSHIRE

JIDDU KRISHNAMURTI: Are all the psychologists, as far as we can understand, really concerned with the future of humanity? Or are they concerned with the human being conforming to the present society? Or going beyond that?

DAVID BOHM: Well, I think that most psychologists evidently want the human being to conform to this society, but I think some are thinking of going beyond that, to transform the consciousness of mankind.

JK: Can the consciousness of mankind be changed through time? That is one of the questions we should discuss.

DB: Yes. We have discussed it already, and I think what came out was that with regard to consciousness, time is not relevant, that it is a kind of illusion. We discussed the illusion of becoming.

JK: Yes, we are saying, aren't we, that the evolution of consciousness is a fallacy.

DB: As through time, yes. Though physical evolution is not.

JK: Can we put it this way, much more simply? There is no psychological evolution, or evolution of the psyche?

DB: Yes, and since the future of humanity depends on the psyche, it seems then that the future of humanity is not going to be determined through actions in time. And that leaves us the question: What will we do?

JK: Now, let's proceed from there. Shouldn't we first distinguish between the brain and the mind?

DB: Well, that distinction has been made, and it is not clear. Now, of course there are several views. One that the mind is just a function of the brain—that is the materialist's view. There is another view which says mind and brain are two different things.

JK: Yes, I think they are two different things.

DB: But there must be ...

JK: ... a contact between the two.

DB: Yes.

JK: A relationship between the two.

DB: We don't necessarily imply any separation of the two.

JK: No. First let's see the brain. I am really not an expert on the structure of the brain and all that kind of thing. But one can see within one, one can observe from one's own activity of the brain, that it is really like a computer which has been programmed and remembers.

DB: Certainly a large part of the activity is that way, but one is not certain that all of it is that way.

JK: No. And it is conditioned.

DB: Yes.

JK: Conditioned by past generations, by society, by the newspapers, by the magazines, by all the activities and pressures from the outside. It is conditioned.

DB: Now, what do you mean by this conditioning?

JK: The brain is programmed. It is made to conform to a certain pattern. It lives entirely on the past, modifying itself with the present and going on.

DB: We have agreed that some of this conditioning is useful and necessary.

JK: Of course.

DB: But the conditioning which determines the self, you know, which determines the . . .

JK: . . . the psyche. Let's call it for the moment the psyche. The self.

DB: The self, the psyche—that conditioning is what you are talking about. That may not only be unnecessary but harmful.

JK: Yes. The emphasis on the psyche, on giving importance to the self, is creating great damage in the world, because it is separative, and therefore it is constantly in conflict, not only within itself but with society, with the family, and so on.

DB: Yes. And it is also in conflict with nature.

JK: With nature, with the whole universe.

DB: We have said that the conflict arose because . . .

JK: . . . of division.

DB: The division arising because thought is limited. Being based on this conditioning, on knowledge and memory, it is limited.

JK: Yes. And experience is limited, therefore knowledge is limited; memory and thought. And the very structure and nature of the psyche is the movement of thought.

DB: Yes.

JK: In time.

DB: Yes. Now, I would like to ask a question. You discussed the movement of thought, but it doesn't seem clear to me what is moving. You see, if I discuss the movement of my hand, that is a real movement. It is clear what is meant. But, now, when we discuss the movement of thought, it seems to me we are discussing something which is a kind of illusion, because you have said that becoming is the movement of thought.

JK: That is what I mean, the movement in becoming.

DB: But you are saying that movement is in some way illusory, aren't you?

JK: Yes, of course.

DB: It is rather like the movement on the screen which is projected from the camera. We say that there are no objects moving across the screen, but the only real movement is the turning of the projector. Now, can we say that there is a real movement in the brain which is projecting all this, which is the conditioning?

JK: That is what we want to find out. Let's discuss that a bit. We both agree, or see, that the brain is conditioned.

DB: We mean really by that that it has been impressed physically and chemically.

JK: And genetically, as well as psychologically.

DB: What is the difference between physically and psychologically?

JK: Psychologically the brain is centred in the self, right?

DB: Yes.

JK: And the constant assertion of the self is the movement, is the conditioning, an illusion.

DB: But there is some real movement happening inside. The brain, for example, is doing something. It has been conditioned physically and chemically. And something is happening physically and chemically when we are thinking of the self.

JK: Are you asking whether the brain and the self are two different things?

DB: No, I am saying that the self is the result of conditioning the brain.

JK: Yes. The self is conditioning the brain.

DB: But does the self exist?

JK: No.

DB: But the conditioning of the brain, as I see it, is the involvement with an illusion which we call the self.

JK: That's right. Can that conditioning be dissipated? That's the whole question.

DB: It really has to be dissipated in some physical and chemical and neurophysiological sense.

JK: Yes.

DB: Now, the first reaction of any scientific person would be that it looks unlikely that we could dissipate it by the sort of thing we are doing. You see, some scientists might feel that maybe we will discover drugs or new genetic changes or deep knowledge of the structure of the brain. In that way we could perhaps hope to do something. I think that idea might be current among some people.

JK: Will that change human behaviour?

DB: Well, why not? I think some people believe it might.

JK: Wait a minute. That is the whole point. It might, which means in the future.

DB: Yes, it would take time to discover all this.

JK: In the meantime, man is going to destroy himself.

DB: Well, they might hope that he will manage to discover it in time. They could also criticize what we are doing, saying what good can it do? You see, it doesn't seem to affect anybody, and certainly not in time to make a big difference.

JK: We two are very clear about it. In what way does it affect humanity?

DB: Will it really affect mankind in time to save . . .

JK: Certainly not. Obviously not.

DB: Then why should we be doing it?

JK: Because this is the right thing to do. Independently. It has nothing to do with reward and punishment.

DB: Nor with goals. You do the right thing even though we don't know what the outcome will be?

JK: That's right.

DB: Are you saying there is no other way?

JK: We are saying there is no other way; that's right.

DB: Well, we should make that clear. For example, some psychologists would feel that by enquiring into this sort of thing, we could bring about an evolutionary transformation of consciousness.

JK: We come back to the point that through time we hope to change consciousness. We question that.

DB: We have questioned that and are saying that, through time, inevitably we are all caught in becoming and illusion, and we will not know what we are doing.

JK: That's right.

DB: Now, could we say that the same thing would hold even for those scientists who are trying to do it physically and chemically or structurally; that they themselves are still caught in this, and through time they are caught in trying to become better?

JK: Yes, the experimentalists and the psychologists and ourselves are all trying to become something.

DB: Yes, though it may not seem obvious at first. It may seem that the scientists are really just disinterested, unbiased observers, working on the problem. But underneath one feels there is the desire to become better on the part of the person who is enquiring in that way.

JK: To become. Of course.

DB: He is not free of that.

JK: That is just it.

DB: And that desire will give rise to self-deception and illusion and so on.

JK: So where are we now? Any form of becoming is an illusion, and becoming implies time, time for the psyche to change. But we are saying that time is not necessary.

DB: That ties up with the other question of the mind and the brain. The brain is an activity in time, as a physical, chemical, complex process.

JK: I think the mind is separate from the brain.

DB: What does separate mean? Are they in contact?

JK: Separate in the sense that the brain is conditioned and the mind is not.

DB: Let's say the mind has a certain independence of the brain. Even if the brain is conditioned . . .

JK: . . . the other is not.

DB: It need not be . . .

JK: . . . conditioned.

DB: On what basis do you say that?

JK: Let's not begin on what basis I say that.

DB: Well, what makes you say it.

JK: As long as the brain is conditioned, it is not free.

DB: Yes.

JK: And the mind is free.

DB: Yes, that is what you are saying. But, you see, the brain not being free means that it is not free to enquire in an unbiased way.

JK: I will go into it. Let's enquire: What is freedom? Freedom to enquire, freedom to investigate. It is only in freedom that there is deep insight.

DB: Yes, that's clear, because if you are not free to enquire, or if you are biased, then you are limited, in an arbitrary way.

JK: So as long as the brain is conditioned, its relationship to the mind is limited.

DB: We have the relationship of the brain to the mind, and also the other way round.

JK: Yes. But the mind being free has a relationship to the brain.

DB: Yes. We are saying the mind is free, in some sense, not subject to the conditioning of the brain.

JK: Yes.

DB: What is the nature of the mind? Is the mind located inside the body, or is it in the brain?

JK: No, it is nothing to do with the body or the brain.

DB: Has it to do with space or time?

JK: Space. Now, wait a minute! It has to do with space and silence. These are the two factors of the . . .

DB: But not time?

JK: Not time. Time belongs to the brain.

DB: You say space and silence; now, what kind of space? It is not the space in which we see life moving.

JK: Space. Let's look round at it the other way. Thought can invent space.

DB: In addition, we have the space that we see. But thought can invent all kinds of space.

JK: And space from here to there.

DB: Yes, the space through which we move is that way.

JK: Space also between two noises, two sounds.

DB: They call that the interval, the interval between two sounds.

JK: Yes, the interval between two noises. Two thoughts. Two notes.

DB: Yes.

JK: Space between two people.

DB: Space between the walls.

JK: And so on. But that kind of space is not the space of the mind.

DB: You say it is not limited?

JK: That's right. But I didn't want to use the word "limited."

DB: But it is implied. That kind of space is not in the nature of being bounded by something.

JK: No, it is not bounded by the psyche.

DB: But is it bounded by anything?

JK: No. So can the brain, with all its cells conditioned, can those cells radically change?

DB: We have often discussed this. It is not certain that all the cells are conditioned. For example, some people think that only some or a small part of the cells are being used and that the others are just inactive, dormant.

JK: Hardly used at all or just touched occasionally.

DB: Just touched occasionally. But those cells that are conditioned, whatever they may be, evidently dominate consciousness now.

JK: Yes. Can those cells be changed?

DB: Yes.

JK: We are saying that they can, through insight; insight being out of time, not the result of remembrance, not an intuition, nor desire, nor hope. It is nothing to do with any time and thought.

DB: Yes. Now, is insight of the mind? Is it of the nature of mind? An activity of mind?

JK: Yes.

DB: Therefore you are saying that mind can act in the matter of the brain.

JK: Yes, we said that earlier.

DB: But, you see, this point, how mind is able to act in matter, is a difficult one.

JK: It is able to act on the brain. For instance, take any crisis or problem. The root meaning of problem is, as you know, "something thrown at you." And we meet it with all the remembrance of the past, with a bias, and so on. And therefore the problem multiplies itself. You may solve one problem, but in the very solution of one particular problem, other problems arise, as happens in politics and so on. Now, to approach the problem, or to have perception of it without any past memories and thoughts interfering or projecting . . .

DB: That implies that perception also is of the mind.

JK: Yes, that's right.

DB: Are you saying that the brain is a kind of instrument of the mind?

JK: An instrument of the mind when the brain is not self-centred.

DB: All the conditioning may be thought of as the brain exciting itself and keeping itself going just from the programme. This occupies all of its capacities.

JK: All our days, yes.

DB: The brain is rather like a radio receiver which can generate its own noise but would not pick up a signal.

JK: Not quite. Let's go into this a little. Experience is always limited. I may blow up that experience into something fantastic and then set up a shop to sell my experience, but that experience is limited. And so knowledge is always limited. And this knowledge is operating in the brain. This knowledge is the brain. And thought is also part of the brain, and thought is limited. So the brain is operating in a very, very small area.

DB: Yes. What prevents it from operating in a broader area? In an unlimited area?

JK: Thought.

DB: But it seems to me the brain is running on its own, from its own programme.

JK: Yes, like a computer.

DB: Now, essentially, what you are asking is that the brain should really be responding to the mind.

JK: It can only respond if it is free from the limited—from thought, which is limited.

DB: So the programme does not then dominate it. You see, we are still going to need that programme.

JK: Of course. We need it for . . .

DB: . . . for many things. But is intelligence from the mind?

JK: Yes, intelligence is the mind.

DB: Is the mind.

JK: We must go into something else. Because compassion is related to intelligence; there is no intelligence without compassion. And compassion can only be when there is love which is completely free from all remembrances, personal jealousies, and so on.

DB: Now, is all that compassion, love, also of the mind?

JK: Of the mind. You cannot be compassionate if you are attached to any particular experience or any particular ideal.

DB: Yes, that is again the programme that is holding us.

JK: Yes. For instance, there are those people who go out to various poverty-ridden countries and work, work, work. And they call that compassion. But they are attached, or tied to a particular form of religious belief, and therefore their action is merely pity or sympathy. It is not compassion.

DB: Yes, well, I understand that we have here two things which can be somewhat independent. There is the brain and the mind, though they make contact. Then we say that intelligence and compassion come from beyond the brain. Now, I would like to go into the question of how they are making contact.

JK: Ah! Contact can only exist between the mind and the brain when the brain is quiet.

DB: Yes, that is the requirement for making it. The brain has got to be quiet.

JK: And that quiet is not a trained quietness. Not a self-conscious, meditative desire for silence. It is a natural outcome of understanding one's own conditioning.

DB: And one can see that if the brain is quiet, you could say it can listen to something deeper?

JK: Deeper—that's right. Then if it is quiet, it is related to the mind. Then the mind can function through the brain.

DB: I think that it would help if we could see with regard to the brain whether it has any activity which is beyond thought. You see, for example, one could ask, is awareness part of the function of the brain?

JK: As long as it is awareness in which there is no choice

DB: I think that may cause difficulty. What is wrong with choice?

JK: Choice means confusion.

DB: That is not obvious.

JK: Of course, you have to choose between two things.

DB: I could choose whether I will buy one thing or another.

JK: Yes, I can choose between this table and that table.

DB: I choose the colours when I buy the table. That need not be confused. If I choose which colour I want, I don't see why that has to be confused.

JK: There is nothing wrong. There is no confusion there.

DB: But it seems to me that the choice about the psyche is where the confusion is.

JK: That's all. We are talking of the psyche that chooses.

DB: That chooses to become.

JK: Yes. Chooses to become. And choice exists where there is confusion.

DB: Are you saying that out of confusion the psyche makes a choice to become one thing or another? Being confused, it tries to become something better?

JK: And choice implies a duality.

DB: But now it seems at first sight that we have another duality that you have introduced, which is the mind and the brain.

JK: No, that is not a duality.

DB: What is the difference?

JK: Let's take a very simple example. Human beings are violent, and nonviolence has been projected by thought. That is the duality—the fact and the non-fact.

DB: You are saying there is a duality between a fact and some mere projection which the mind makes.

JK: The ideal and the fact.

DB: The ideal is unreal, and the fact is real.

JK: That's it. The ideal is not actual.

DB: Yes. Now, you are saying the division of those two is duality. Why do you give it that name?

JK: Because they are divided.

DB: Well, at least they appear to be divided.

JK: Divided, and we are struggling. For instance, all the totalitarian communist ideals, and the democratic ideals, are the outcome of thought which is limited, and this is creating havoc in the world.

DB: So there is a division which has been brought in. But I think we were discussing in terms of dividing something which cannot be divided. Of trying to divide the psyche.

JK: That's right. Violence cannot be divided into nonviolence.

DB: And the psyche cannot be divided into violence and nonviolence. Right?

JK: It is what it is.

DB: It is what it is. So if it is violent, it can't be divided into a violent and a nonviolent part.

JK: So can we remain with *what is,* not with *what should be, what must be,* not invent ideals and so on?

DB: Yes, but could we return to the question of the mind and the brain? Now, we are saying that is not a division.

JK: Oh, no, that is not a division.

DB: They are in contact, is that right?

JK: We said there is contact between the mind and the brain when the brain is silent and has space.

DB: So we are saying that although they are in contact and not divided at all, the mind can still have a certain independence of the conditioning of the brain.

JK: Now, let's be careful! Suppose my brain is conditioned, for example, programmed as a Hindu, and my whole life and action are conditioned by the idea that I am a Hindu. Mind obviously has no relationship with that conditioning.

DB: You are using the word "mind," not "my mind."

JK: Mind. It is not "mine."

DB: It is universal or general.

JK: Yes. And it is not "my brain" either.

DB: No, but there is a particular brain, this brain or that brain. Would you say that there is a particular mind?

JK: No.

DB: That is an important difference. You are saying mind is really universal.

JK: Mind is universal—if you can use that ugly word.

DB: Unlimited and undivided.

JK: It is unpolluted, not polluted by thought.

DB: But I think for most people there will be difficulty in saying how we know anything about this mind. I only know of my mind is the first feeling, right?

JK: You cannot call it your mind. You only have your brain, which is conditioned. You can't say, "It is my mind."

DB: But whatever is going on inside I feel is mine, and it is very different from what is going on inside somebody else.

JK: No, I question whether it is different.

DB: At least it seems different.

JK: Yes. I question whether it is different, what is going on inside me as a human being and inside you as another human being. We both go through all kinds of problems: suffering, fear, anxiety, loneliness, and so on. We have our dogmas, beliefs, superstitions. And everybody has this.

DB: Well, we'll say it is all very similar, but it still seems as if each one of us is isolated from the other.

JK: By thought. My thought has created the belief that I am different from you, because my body is different from yours, my face is different from yours. We extend the same thing into the psychological area.

DB: But, now, if we said all right that division is an illusion, perhaps?

JK: No, not perhaps! It is.

DB: It is an illusion. All right. Although it is not obvious when a person first looks at it.

JK: Of course.

DB: In reality even brain is not divided, because we are saying that we are all not only basically similar but really connected, right? And then we say beyond all that is mind, which has no division at all.

JK: It is unconditioned.

DB: Yes, it would almost seem to imply, then, that insofar as a person feels he is a separate being, he has very little contact with mind, right?

JK: Absolutely. Quite right. That is what we said.

DB: No mind.

JK: That is why it is very important to understand not the mind but my conditioning. And whether my conditioning, human conditioning, can ever be dissolved. That is the real issue.

DB: Yes, I think we still want to understand the meaning of what is being said. You see, we have a mind that is universal; that is in some kind of space, you say, or is it its own space?

JK: It is not in me or in my brain.

DB: But it has a space.

JK: It is, it lives in, space and silence.

DB: It lives in a space and silence, but it is the space of the mind. It is not a space like this space?

JK: No. That is why we said space is not invented by thought.

DB: Yes, now is it possible, then, to perceive this space when the mind is silent, to be in contact with it?

JK: Not perceive. Let's see. You are asking whether the mind can be perceived by the brain.

DB: Or at least the brain somehow be aware . . . an awareness, a sense.

JK: We are saying, yes. Through meditation. You may not like to use that word.

DB: I don't mind.

JK: You see, the difficulty is that when you use the word "meditation" it is generally understood that there is always a meditator meditating. Meditation is really an unconscious process, not a conscious process.

DB: But how can you say that meditation takes place if it is unconscious?

JK: It is taking place when the brain is quiet.

DB: You mean by consciousness all the movement of thought? Feeling, desire, will, and all that goes with it?

JK: Yes.

DB: There is a kind of awareness still, isn't there?

JK: Oh, yes. It depends what you call awareness. Awareness of what?

DB: Possibly awareness of something deeper. I don't know.

JK: Again, when you use the word "deeper," it is a measurement. I wouldn't use that.

DB: Well, let's not use that. But, you see, there is a kind of unconsciousness which we are simply not aware of at all. A person may be unconscious of some of his problems, conflicts.

JK: Let's go at it a bit more. If I do something consciously, it is the activity of thought.

DB: Yes, it is thought reflecting on itself.

JK: Right. It is the activity of thought. Now, if I consciously meditate, practise, do all that, which I call nonsense, then I am making the brain conform to another series of patterns.

DB: Yes, it is more becoming.

JK: More becoming—that's right.

DB: You are trying to become better.

JK: There is no illumination by becoming. One can't be illumined, if I can use that word, by saying that one is going to be a cheap guru.

DB: But it seems very difficult to communicate something which is not conscious.

JK: That's it. That's the difficulty.

DB: It is not just being knocked out. If a person is unconscious, he is knocked out, but you don't mean that.

JK: Of course not!

DB: Or under anaesthetic or . . .

JK: No. Let's put it this way: Conscious meditation, conscious activity to control thought, to free oneself from conditioning, is not freedom.

DB: Yes. I think that is clear, but it becomes very unclear how to communicate something else.

JK: Wait a minute. Say that you want to tell me what lies beyond thought.

DB: Or when thought is silent.

JK: Quite, silent. What words would you use?

DB: Well, I suggested the word "awareness." What about the word "attention"?

JK: Attention for me is better. Would you say, in attention there is no centre as the me?

DB: Well, not in the kind of attention you are discussing. There is the usual kind, where we pay attention because of what interests us.

JK: Attention is not concentration.

DB: We are discussing a kind of attention without the "me" present and which is not the activity of conditioning.

JK: Not the activity of thought. In attention, thought has no place.

DB. Yes, but could we say more? What do you mean by attention? Now, would the derivation of the word be of any use? It means stretching the mind. Would that help?

JK: No. Would it help if we say concentration is not attention? Effort is not attention. When I make an effort to attend, it is not attention. Attention can come into being only when the self is not.

DB: Yes, but that is going to get us in a circle, because we are starting when the self is.

JK: No, I used the word carefully. Meditation means measure.

DB: Yes.

JK: As long as there is measurement, which is becoming, there is no meditation. Let's put it that way.

DB: Yes. We can discuss when there is not meditation.

JK: That's right. Through negation the other is.

DB: Because if we succeed in negating the whole activity of what is not meditation, the meditation will be there.

JK: That's right.

DB: That which is not meditation but which we think is meditation.

JK: Yes, that is very clear. As long as there is measurement, which is the becoming, which is the process of thought, meditation or silence cannot be.

DB: Is this undirected attention of the mind?

JK: Attention is of the mind.

DB: And it contacts the brain, doesn't it?

JK: Yes. As long as the brain is silent, the other has contact.

DB: That is, this true attention has contact with the brain, when the brain is silent.

JK: Silent, and has space.

DB: What is space?

JK: The brain has no space now, because it is concerned with itself, it is programmed, it is self-centred, and it is limited.

DB: Yes. The mind is in its space. Now, does the brain have its space too? Limited space?

JK: Of course. Thought has a limited space.

DB: But when thought is absent, does the brain have its space?

JK: Yes. The brain has space.

DB: Unlimited?

JK: No. It is only mind that has unlimited space. My brain can be quiet over a problem which I have thought about, and I suddenly say, "Well, I won't think anymore about it," and there is a certain amount of space. In that space you solve the problem.

DB: Now, if the brain is silent, if it is not thinking of a problem, then still the space is limited, but it is open to . . .

JK: . . . to the other.

DB: To the attention. Would you say that through attention, or in attention, the mind is contacting the brain?

JK: When the brain is not inattentive.

DB: So what happens to the brain?

JK: What happens to the brain? Which is to act. Let's get it clear. We said intelligence is born out of compassion and love. That intelligence operates when the brain is quiet.

DB: Yes. Does it operate through attention?

JK: Of course.

DB: So attention seems to be the contact.

JK: Naturally. We said too that attention can only be when the self is not.

DB: Now, you say that love and compassion are the ground, and out of this comes the intelligence, through attention.

JK: Yes, it functions through the brain.

DB: So there are two questions: One is the nature of this intelligence, and the second is what does it do to the brain?

JK: Yes. Let's see. We must again approach it negatively. Love is not jealousy and all that. Love is not personal, but it can be personal.

DB: Then it is not what you are talking about.

JK: Love is not *my* country, *your* country, or "I love *my* God." It is not that.

DB: Well, if it is from universal mind . . .

JK: That is why I say love has no relationship to thought.

DB: Yes, and it does not start, does not originate in the particular brain.

JK: When there is that love, out of it there is compassion and there is intelligence.

DB: Is this intelligence able to understand deeply?

JK: No, not "understand."

DB: What does it do? Does it perceive?

JK: Through perception it acts.

DB: Perception of what?

JK: Now, let's discuss perception. There can be perception only when it is not tinged by thought. When there is no interference from the movement of thought, there is perception, which is direct insight into a problem or into human complexities.

DB: Yes. Now, this perception originates in the mind?

JK: Does the perception originate in the mind? Yes. When the brain is quiet.

DB: We used the words "perception" and "intelligence." Now, how are they related, or what is their difference?

JK: The difference between perception and intelligence?

DB: Yes.

JK: None.

DB: So we can say intelligence is perception.

JK: Yes, that's right.

DB: Intelligence is perception of *what is*, right? And through attention there is contact.

JK: Let's take a problem. Then it is easier to understand. Take the problem of suffering. Human beings have suffered endlessly, through wars, through physical disease, and through wrong relationship with each other. Now, can that end?

DB: I would say the difficulty of ending that is that it is on the programme. We are conditioned to this whole thing, physically and chemically.

JK: Yes. And that has been going on for centuries.

DB: So it is very deep.

JK: Very, very deep. Now, can that suffering end?

DB: It cannot end by an action of the brain.

JK: By thought.

DB: Because the brain is caught in suffering, and it cannot take an action to end its own suffering.

JK: Of course it cannot. That is why thought cannot end it. Thought has created it.

DB: Yes, thought has created it, and anyway it is unable to get hold of it.

JK: Thought has created the wars, the misery, the confusion. And thought has become prominent in human relationship.

DB: Yes, but I think people might agree with that and still think that just as thought can do bad things, it can do good things.

JK: No, thought cannot do good or bad. It is thought, limited.

DB: Thought cannot get hold of this suffering. That is, this suffering being in the physical and chemical conditioning of the brain, thought has no way of even knowing what it is.

JK: I mean, I lose my son and I am . . .

DB: Yes, but by thinking, I don't know what is going on inside me. I can't change the suffering inside because thinking will not show me what it is. Now, you are saying intelligence, perception, does.

JK: But we are asking, can suffering end? That is the problem.

DB: Yes, and it is clear that thinking cannot end it.

JK: Thought cannot do it. That is the point. If I have an insight into it . . .

DB: Yes. Now, this insight will be through the action of the mind, through intelligence and attention.

JK: When there is that insight, intelligence wipes away suffering.

DB: Now, you are saying, therefore, that there is a contact from mind to matter which removes the whole physical, chemical structure that keeps us going on with suffering.

JK: That's right. In that ending there is a mutation in the brain cells. We have discussed this some years ago.

DB: Yes, and that mutation just wipes out the whole structure that makes you suffer.

JK: That's right. Therefore it is as if I have been going along a certain tradition; I suddenly change that tradition and there is a change in the whole brain, which has been going north. Now it goes east.

DB: Of course this is a radical notion from the point of view of traditional ideas in science, because if we accept that mind is different from matter, then people would find it hard to say that mind would actually . . .

JK: Would you put it that mind is pure energy?

DB: Well, we could put it that way, but matter is energy too.

JK: But matter is limited; thought is limited.

DB: But we are saying that the pure energy of mind is able to reach into the limited energy of matter?

JK: Yes, that's right. And change the limitation.

DB: Remove some of the limitation.

JK: When there is a deep issue, problem, or challenge which you are facing.

DB: We could also add that all the traditional ways of trying to do this cannot work.

JK: They haven't worked.

DB: Well, that is not enough. We have to say, because people still might hope it could, that it cannot, actually.

JK: It cannot.

DB: Because thought cannot get at its own physical, chemical basis in the cells and do anything about those cells.

JK: Yes. Thought cannot bring about a change in itself.

DB: And yet practically everything that mankind has been trying to do is based on thought. There is a limited area, of course, where that is all right, but we cannot do anything about the future of humanity from that usual approach.

JK: When one listens to the politicians, who are so very active in the world, they are creating problem after problem, and to them thought, ideals, are the most important things.

DB: Generally speaking nobody knows of anything else.

JK: Exactly. We are saying that the old instrument which is thought is worn out, except in certain areas.

DB: It never was adequate, except in those areas.

JK: Of course.

DB: And as far as history goes, man has always been in trouble.

JK: Man has always been in trouble, in turmoil, in fear. And facing all the confusion of the world, can there be a solution to all this?

DB: That comes back to the question I would like to repeat. It seems there are a few people who are talking about it and think perhaps they know, or perhaps they meditate and so on. But how is that going to affect this vast current of mankind?

JK: Probably very little. But why will it affect this? It might, or it might not. But then one puts that question: What is the use of it?

DB: Yes, that's the point. I think there is an instinctive feeling that makes one put the question.

JK: But I think that is the wrong question.

DB: You see, the first instinct is to say, "What can we do to stop this tremendous catastrophe?"

JK: Yes. But if each one of us, whoever listens, sees the truth that thought, in its activity both externally and inwardly, has created a terrible mess, great suffering, then one must inevitably ask is there an ending to all this? If thought cannot end it, what will?

DB: Yes.

JK: What is the new instrument that will put an end to all this misery? You see, there is a new instrument which is the mind, which is intelligence. But the difficulty is also that people won't listen to all this. Both the scientists and the ordinary laymen like us, have come to definite conclusions, and they won't listen.

DB: Yes. Well, that is what I had in mind when I said that a few people don't seem to have much effect.

JK: Of course. I think, after all, few people have changed the world, whether good or bad, but that is not the point. Hitler and the communists have changed it, but they have gone to the same pattern again. Physical revolution has never psychologically changed the human state.

DB: Do you think it is possible that a certain number of brains coming in contact with mind in this way will be able to have an effect on mankind, which is beyond just the immediate, obvious effect of their communication?

JK: Yes, that's right. But how do you convey—I have often thought about it—this subtle and very complex issue to a person who is steeped in tradition, who is conditioned and won't even take time to listen, to consider?

DB: Well, that is the question. You see, you could say that this conditioning cannot be absolute, cannot be an absolute block, or else there would be no way out at all. But the conditioning may be thought to have some sort of permeability.

JK: I mean, after all, the pope won't listen to us, but the pope has tremendous influence.

DB: Is it possible that every person has something he can listen to, if it could be found?

JK: If he has a little patience. Who will listen? The politicians won't listen. The idealists won't listen. The totalitarians won't listen. The deeply steeped religious people won't listen. So perhaps a so-called ignorant person, not highly educated or conditioned in his professional career or by money, the poor man who says, "I am suffering. Please let's end that."

DB: But he doesn't listen either, you see. He wants to get a job.

JK: Of course. He says, "Feed me first." We have been through all this for the last sixty years. The poor man won't listen, the rich man won't listen, the learned won't listen, and the deeply dogmatic religious believers don't listen. So perhaps it is like a wave in the world; it might catch somebody. I think it is a wrong question to say, does it affect?

DB: Yes. All right. We will say that that brings in time, and that is becoming. It brings in the psyche in the process of becoming again.

JK: Yes. But . . . if you say . . . it must affect mankind . . .

DB: Are you proposing that it affects mankind through the mind directly rather than through . . .

JK: Yes. It may not show immediately in action.

DB: You are taking very seriously what you said, that the mind is universal and is not located in our ordinary space, is not separate.

JK: Yes, but there is a danger in saying this, that the mind is universal. That is what some people say of the mind, and that has become a tradition.

DB: One can turn it into an idea, of course.

JK: That is just the danger of it; that is what I am saying.

DB: Yes. But really the question is we have to come directly in contact with this to make it real. Right?

JK: That's it. We can only come into contact with it when the self is not. To put it very simply, when the self is not, there is beauty, silence, space. Then that intelligence, which is born of compassion, operates through the brain. It is very simple.

DB: Yes. Would it be worth discussing the self, since the self is widely active?

JK: I know. That is our long tradition of many, many centuries.

DB: Is there some aspect of meditation which can be helpful here when the self is acting? You see, suppose a person says, "All right, I am caught in the self, but I want to get out. But I want to know what I shall do?"

JK: Ah, you see, that is . . .

DB: I won't use the words "What shall I do?" but what do you say?

JK: That is very simple. Is the observer different from the observed?

DB: Well, suppose we say yes, it appears to be different. Then what?

JK: Is that an idea or an actuality?

DB: What do you mean?

JK: Actuality is when there is no division between the thinker and the thought.

DB: But suppose I say ordinarily one feels that the observer is different from the observed. We begin there.

JK: We begin there. I'll show you. Look at it. Are you different from your anger, from your envy, from your suffering? You are not.

DB: At first sight it appears that I am, that I might try to control it.

JK: You are that.

DB: Yes, but how will I see that I am that?

JK: You are your name. You are your form, your body. You are the reactions and actions. You are the belief, the fear, the suffering and pleasure. You are all that.

DB: But the first experience is that I am here first and that those are properties of me; they are my qualities which I can either have or not have. I might be angry or not angry; I might have this belief or that belief.

JK: Contradictory. You are all that.

DB: But, you see, it is not obvious. When you say I am that, do you mean that I am that and cannot be otherwise?

JK: At present you are that. It can be totally otherwise.

DB: All right. So I am all that. I usually say that I am looking at these qualities. I feel that I am not angry but an unbiased observer looking at anger. And you are telling me that this unbiased observer is the same as the anger he is looking at?

JK: Of course. Just as I analyze myself, and the analyzer is the analyzed.

DB: Yes. He is biased by what he analyzes.

JK: Yes.

DB: So if I watch anger for a while, I can see that I am very biased by the anger. So at some stage I say that I am one with that anger?

JK: No, not "I am one with it"; "I *am* it."

DB: That anger and I are the same?

JK: Yes. The observer is the observed. And when there is that actuality, you have really eliminated conflict altogether. Conflict exists when I am separate from my quality.

DB: Yes. That is because if I believe myself to be separate, then I can try to change it, but since I am that, it is trying to change itself and remain itself at the same time.

JK: Yes, that's right. But when the quality is me, the division has ended. Right?

DB: When I see that the quality is me, then there is no point in trying to change it.

JK: No. What happened before when the quality is not me, then in that there is conflict, either suppression or escape and so on, which is a wastage of energy. When that quality is me, all that energy which has been wasted is there to look, to observe.

DB: But why does it make such a difference to have that quality being me?

JK: It makes a difference when there is no division between the quality and me.

DB: Yes, when there is no perception of a difference . . .

JK: That's right. Put it round differently.

DB: The mind does not try to fight itself.

JK: Yes, yes. *It is so.*

DB: If there is an illusion of a difference, the mind must be compelled to fight against itself.

JK: The brain.

DB: The brain fights against itself.

JK: That's right.

DB: On the other hand, when there is no illusion of a difference, the brain just stops fighting.

JK: And therefore you have tremendous energy.

DB: The brain's natural energy is released?

JK: Yes. And energy means attention.

DB: The energy of the brain allows for attention.

JK: For that thing to dissolve.

DB: Yes, but wait a minute. We said before that attention was a contact of the mind and the brain.

JK: Yes.

DB: The brain must be in a state of high energy to allow that contact.

JK: That's right.

DB: I mean a brain which is low energy cannot allow that contact.

JK: Of course not. But most of us are low energy because we are so conditioned.

DB: Well, essentially you are saying that this is the way to start.

JK: Yes, start simply. Start with *what is, what I am.* Self-knowledge is so important. It is not an accumulated process of knowledge, which one then looks at; it is a constant learning about oneself.

DB: If you call it self-knowledge, then it is not knowledge of the kind we talked about before, which is conditioning.

JK: That's right. Knowledge conditions.

DB: But you are saying that self-knowledge of this kind is not conditioning. But why do you call it knowledge? Is it a different kind of knowledge?

JK: Yes. Accumulated knowledge conditions.

DB: Yes, but now you have this self-knowledge.

JK: Which is to know and to comprehend oneself. To understand oneself is such a subtle, complex thing. It is living.

DB: Essentially knowing yourself in the very moment in which things are happening.

JK: Yes, to know what is happening.

DB: Rather than store it up in memory.

JK: Of course. Through reactions, I begin to discover what I am.

JIDDU KRISHNAMURTI *(1895–1986),* who had been taken up by the Theosophical Society as a World Teacher at the age of thirteen, firmly renounced any messianic role in 1929. Until his death in 1986, he travelled the world, constantly giving public talks and interviews, denying the notion of authority in religious matters, and inviting his listeners to explore for themselves the possibility of freeing human consciousness from self-imposed limitations.

DAVID BOHM *(1917–1992)* was one of the most significant theoretical physicists of the twentieth century and a Fellow of the Royal Society. He was the author of many scientific works, ranging from the fundamental textbook *Quantum Theory* to *Causation and Chance in Modern Physics, The Special Theory of Relativity,* and the innovative *Wholeness and the Implicate Order.* Like many leading physicists, he was intrigued by the role of the observer in the universe, and after reading Krishnamurti's *The First and Last Freedom* in 1961, he decided to contact its author. This began a series of discussions and a friendship that lasted for the next two decades. While continuing with groundbreaking research in physics, his interest in Krishnamurti's insights never flagged; it was characteristic of Bohm's intensely enquiring mind that the day before his death he remarked to his wife, Saral, after viewing a video of himself in dialogue with Krishnamurti, "We should have gone further."

DAVID SKITT is a former trustee of the Krishnamurti Foundation Trust in Britain and the editor of a number of books of Krishnamurti's talks and dialogues. He also worked as an editor for many years in the European Space Agency.